AF553367

TEXTBOOK OF PHOTOCHEMISTRY

TEXTBOOK OF PHOTOCHEMISTRY

By

Dr. Shubhrata R. Mishra

Deptt. of Botany
Vikram University
Ujjain (M.P.)
(India)

DISCOVERY PUBLISHING HOUSE PVT. LTD.
NEW DELHI-110 002

Published by:
Tilak Wasan

DISCOVERY PUBLISHING HOUSE PVT. LTD.
4831/24, Ansari Road, Prahlad Street
Darya Ganj, New Delhi-110002 (India)
Phone: +91-11-23279245, 43764432
Fax: +91-11-23253475
E-mail: parul.wasan@gmail.com
discoverypublishinghouse@gmail.com
info@discoverypublishinggroup.com
web: www.discoverypublishinggroup.com

First Edition: **2011**
ISBN: 978-81-8356-846-3

Textbook of Photochemistry

Printed at:
Shree Balaji Art Press
Delhi

PREFACE

Like most scientific disciplines, photochemistry utilizes the SI or metric measurement system. Important units and constants that show up regularly include the meter (and variants such as centimeter, millimeter, micrometer, nanometer, etc.), seconds, hertz, joules, moles, the gas constant R, and the Boltzmann constant. These units and constants are also integral to the field of physical chemistry.

The first law of photochemistry, known as the Grotthuss-Draper law (for chemists Theodor Grotthuss and John W. Draper), states that light must be absorbed by a chemical substance in order for a photochemical reaction to take place.

The second law of photochemistry, the Stark-Einstein law, states that for each photon of light absorbed by a chemical system, only one molecule is activated for a photochemical reaction. This is also known as the photoequivalence law and was derived by Albert Einstein at the time when the quantum (photon) theory of light was being developed.

Photochemistry may also be introduced to laymen as a reaction that proceeds with the absorption of light. Normally a reaction (not just a photochemical reaction) occurs when a molecule gains the necessary activation energy to undergo change. A simple example can be the combustion of gasoline (a hydrocarbon) into carbon dioxide and water. This is a chemical reaction where one or more molecules/chemical species are converted into others. For this reaction to take place activation energy should be supplied. The activation energy is provided in the form of heat or a spark. In case of photochemical reactions light provides the activation energy.

Author

CONTENTS

CHAPTER 1 INTRODUCTION

Photochemistry, a sub-discipline of chemistry, is the study of the interactions between atoms, small molecules, and light (or electromagnetic radiation). The pillars of photochemistry are UV/VIS spectroscopy, photochemical reactions in organic chemistry and photosynthesis in biochemistry.

Scientific Background

Like most scientific disciplines, photochemistry utilizes the SI or metric measurement system. Important units and constants that show up regularly include the meter (and variants such as centimeter, millimeter, micrometer, nanometer, etc.), seconds, hertz, joules, moles, the gas constant R, and the Boltzmann constant. These units and constants are also integral to the field of physical chemistry.

The first law of photochemistry, known as the Grotthuss-Draper law (for chemists Theodor Grotthuss and John W. Draper), states that light must be absorbed by a chemical substance in order for a photochemical reaction to take place.

The second law of photochemistry, the Stark-Einstein law, states that for each photon of light absorbed by a chemical system, only one molecule is activated for a photochemical reaction. This is also known as the photoequivalence law and was derived by Albert Einstein at the time when the quantum (photon) theory of light was being developed.

Photochemistry may also be introduced to laymen as a reaction that proceeds with the absorption of light. Normally a reaction (not just a photochemical reaction) occurs when a molecule gains the necessary activation energy to undergo change. A simple example can be the combustion of gasoline (a hydrocarbon) into carbon dioxide and water. This is a chemical reaction where one or more molecules/chemical species are converted into others. For this reaction to take place activation energy should be supplied. The activation energy is provided in the form of heat or a spark. In case of photochemical reactions light provides the activation energy.

The absorption of a photon of light by a reactant molecule may also permit a reaction to occur not just by bringing the molecule to the necessary activation energy, but also by changing the symmetry of the molecule's electronic configuration, enabling an otherwise inaccessible reaction path, as described by the Woodward-Hoffmann selection rules. A 2+2 cycloaddition reaction is one example of a pericyclic reaction that can be analyzed using these rules or by the related frontier molecular orbital theory.

Spectral Regions

Photochemists typically work in only a few sections of the electromagnetic spectrum. Some of the most widely used sections, and their wavelengths, are the following:

- Ultraviolet: 100–400 nm
- Visible Light: 400–700 nm
- Near infrared: 700–2500 nm

- Mid infrared: 2500 - 25000 nm
- Far infrared: 25–1000 μm

Applications

There are important processes based in the photochemistry principles. One case is photosynthesis, which some plants use light to create glucose in their chloroplasts to contribute to cell metabolism. The glucose is used by the plant's mitochondria to produce adenosine triphosphate. Medicine bottles are made with darkened glass to prevent the medicine itself from reacting chemically with light. In fireflies, an enzyme in the abdomen works to produce bioluminescence. The mercaptans or thiols produced by Chevron Phillips Chemical Company are produced by photochemical addition of hydrogen sulfide (H2S) to alfa olefins. Among their many uses as a chemical reagent these mercaptans are used to provide a distinctive odor (an odorant) to otherwise odorless natural gas. Many polymerizations are started by photoinitiatiors which decompose upon absorbing light to produce the necessary free radicals for Radical polymerization.

When the sodium atom absorbs a photon it is said to be excited. After a short period of time, the excited state sodium atom emits a photon of 589 nm light and falls back to the ground state:

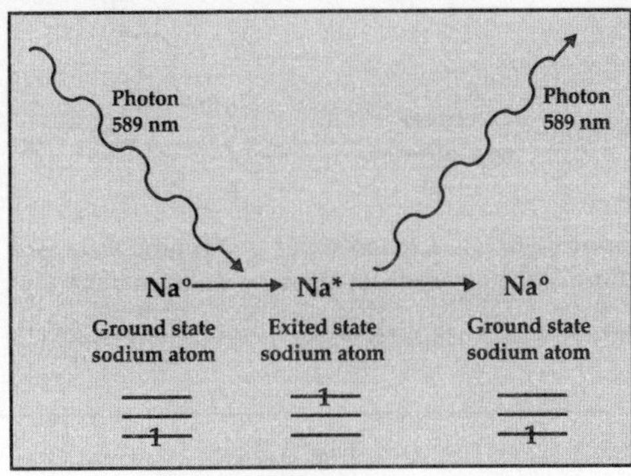

Fig. 1.1

The atom can be excited by a flame, and this is the basis of the well known "flame test" of Group 1 and 2 metal salts (LiCl, NaBr, $CaCl_2$, $SrCl_2$, etc.) and atomic emission spectroscopy.:

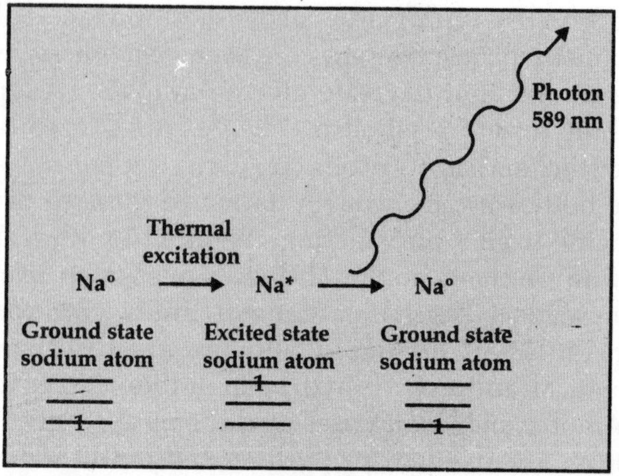

Fig. 1.2

Or the sodium atoms can be electrically excited and this is the basis of energy efficient orange sodium street lights:

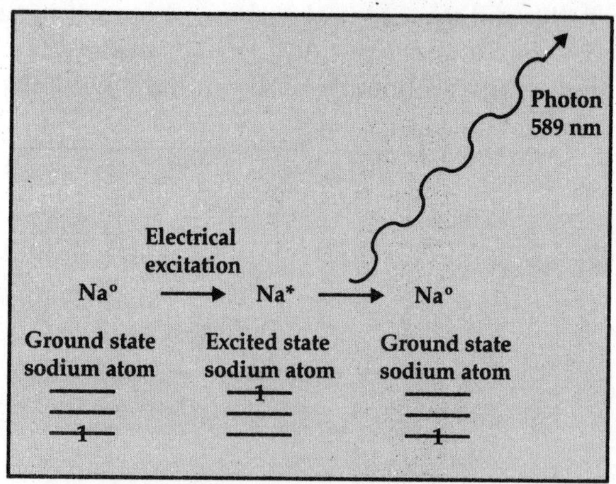

Fig. 1.3

The Electromagnetic Spectrum

Photons have an energy which is dependent upon the wavelength of the light. The rule is:

Long wavelength light = low energy

Short wavelength light = high energy

When an opera diva shatters a glass with her singing voice, the glass absorbs resonant sound energy. To be "resonant" the frequency (or wavelength or energy) of the sound must exactly match that of the glass. And so it is with electromagnetic radiation and chemical species.

Spectroscopy is the study of absorption and emission of photons by atoms, ions and molecules. Studies reveal details of atomic, molecular and ionic structure. However, the revealed world is often strange because it is governed by quantum mechanics and subtle selection rules.

Photons of electromagnetic radiation of many wavelengths (or energies) interact with chemical species in a variety of ways.

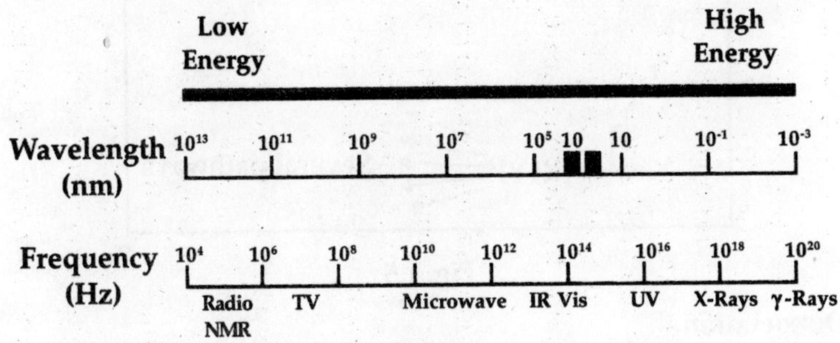

Fig. 1.4

- In a strong magnetic field many atomic nuclei are made to resonate by radio waves. This is the nuclear magnetic resonance (NMR) effect.
- In the gas phase molecules are made to rotate by microwaves.

- Infrared radiation causes molecular bonds to stretch and vibrate. Infrared spectroscopy is often called "vibrational spectroscopy".
- Visible radiation induces low energy electronic transitions in atom's and molecules.
- Ultraviolet radiation causes high energy electronic transitions in molecules. The resulting excited states may relax via bond breakage. Traditionally most photo reaction chemistry has been performed with UV lamp sources. However, laser technology is extending the wavelength range of photoactivation.
- X-rays excite and eject inner shell electrons. This causes widespread ionisation and bond fragmentation.
- Gamma-Rays are highly penetrating ionising radiation which are generated during transitions in atomic nuclei.

Photochemical Reaction Processes

A photoexcited species, [X-Y]*, can relax (react) via a variety of pathways.

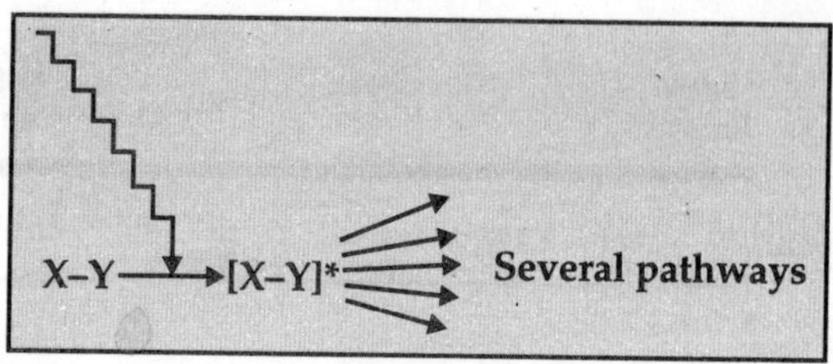

Fig. 1.5

Dissociation

$[X\text{-}Y]^* \rightarrow X\text{-}Y$

$[Cl\text{-}Cl]^* \rightarrow Cl^\bullet + Cl^\bullet$

$[NO_2]^* \rightarrow NO + O$

The excited state species may fragment to a pair of radicals or, in the case of nitrogen dioxide to nitric oxide and oxene.

Reactions of this type, photolysis, are important in gas phase and atmospheric chemistry. Many species can be induced to photodissociate.

The photoexcited state may undergo reactions unavailable to the ground state species. For example, a photoexcited ketone can undergo a 2+2 cycloaddition with an alkene to give an oxetane (a cyclic ether with a four membered ring), a Paterno-Büchi reaction.

Direct Reaction

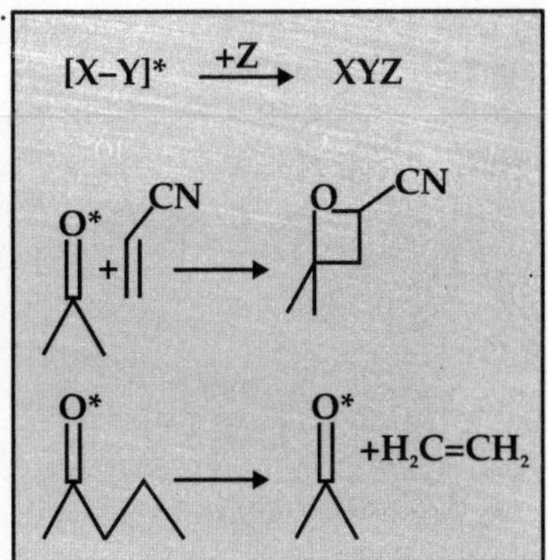

Fig. 1.6

Woodward and Hoffmann showed that photo initiated 2+2 cycloadditions were complementary to the more common 2+4 thermal cycloadditions, such as Diels-Alder cycloaddition.

Photoexcited species may also be able to fragment. For example, 2-pentanone (also called methyl propyl ketone) can be photoexcited to a species which eliminates ethene (ethylene) to give propanone (acetone).

Isomerisation

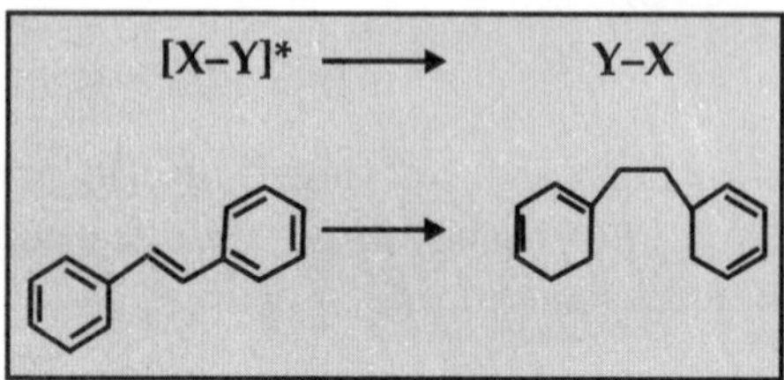

Fig. 1.7

Photoexcited species may undergo isomerisation. For example, trans-stilbene can be photoexcited to a state that allows free rotation around the alkene double bond. The photoexcited species is able to relax back to the ground state at any time, and if this happens when the excited stilbene is in the cis conformation, ground state cis-stilbene will be formed.

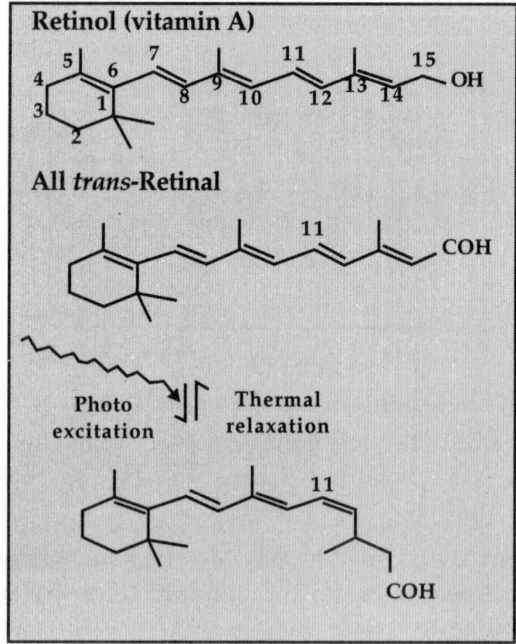

Fig. 1.8

This is photo induced trans-to-cis isomerism exploited in the visual system where the conversion of all-trans-retinal to 11-cis-retanal is the main photon detector. With time, the 11-cis-retanal thermally relaxes back to the all-trans-retinal configuration. Retinal is derived from retinol, also known as vitamin A.

An excited state species can transfer energy to a ground state species. This process is used to produce singlet oxygen. A dye, usually rose bengal, is photoexcited with UV light, here. The photoexcited is able to transfer energy to triplet oxygen, which is converted to singlet oxygen. Singlet oxygen has a different reactivity spectrum compared with the triplet species.

The energy state may undergo intersystem crossing and relax in an intramolecular manner. For example, the S1(excited singlet) state of naphthalene can convert into the T1 (excited triplet) state.

Energy Transfer by Collision

$$[X\text{–}Y]^* \xrightarrow{+Z} [Z]^* + X\text{-}Y$$

$$[\text{Rose Bengal}]^* + \underset{\text{Triplet}}{O_2} \longrightarrow \underset{\text{Singlet}}{O_2} + \text{Rose Bengal}$$

Fig. 1.9

Energy Transfer (Internal)

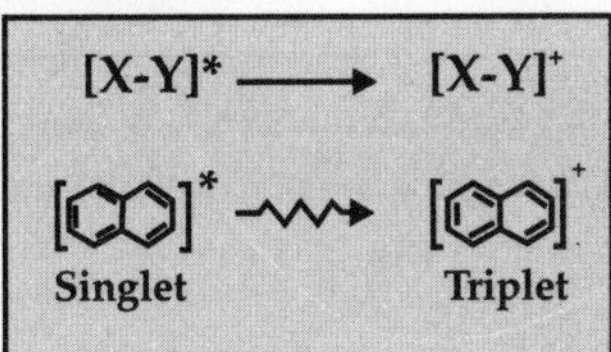

Fig. 1.10

Excited state species can emit light in a process called luminescence. If the process is quick, the process is known as fluorescence and if it is delayed it is known as phosphorescence.

Luminescence

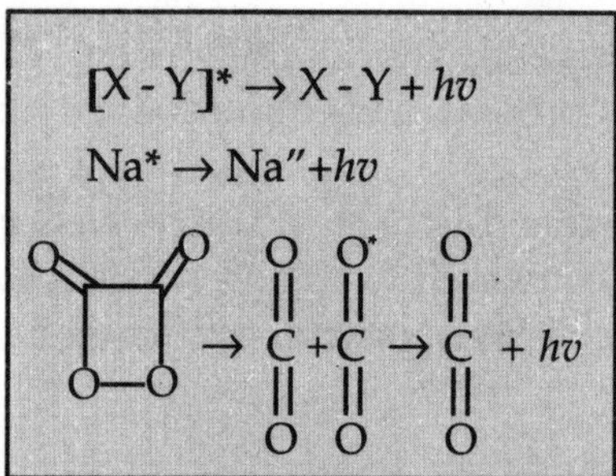

Fig. 1.11

If the excited state is produced chemically, the system can be used to produce "chemical light":

> The four membered ring system shown above (oxalic per-anhydride) is a carbon dioxide dimer. The ring strained species is able to undergo a retrocycloaddition to produce two molecules of carbon dioxide, one of which is highly excited. This excited state species quickly emits a photon of UV light. If there is a fluorescent dye present, this photon of UV light can be converted into a photon of visible light. This chemistry in used in the popular "light sticks".

Quenching

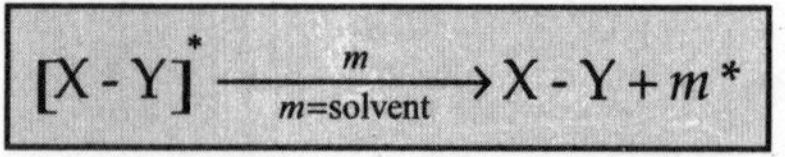

Fig. 1.12

In the liquid state, the excited state species may be quenched in which the energy of the excited state species is converted into vibrational energy (heat).

Photo Ionisation

$$[X\text{-}Y]^* \rightarrow [X\text{-}Y]^{+} + e^-$$

$$NO + \underset{134.3nm}{\overset{hv}{}} \rightarrow [NO]^* \rightarrow NO^+ + e^-$$

Fig. 1.13

Photoionisation processes are of great importance high in the atmosphere where pressures are low and short wavelength UV radiation from the sun has a high flux.

CHAPTER 2
PHOTON

In physics, the photon is an elementary particle, the quantum of the electromagnetic field and the basic unit of light and all other forms of electromagnetic radiation. It is also the force carrier for the electromagnetic force. This force's easily visible human-scale effects and applications, from sunlight to radiotelephones, are due to the fact that the photon has no mass and thus can produce interactions at long distances. Like all elementary particles, the photon is governed by quantum mechanics and so exhibits wave-particle duality: that is, it exhibits both wave and particle properties. For example, a single photon may undergo refraction by a lens or exhibit wave interference, but also act as a particle giving a definite result when its location is measured.

The modern concept of the photon was developed gradually by Albert Einstein to explain experimental observations that did not fit the classical wave model of light. In particular, the photon model accounted for the frequency dependence of light's energy, and explained the ability of matter and radiation to be in thermal equilibrium. Other physicists sought to explain these anomalous observations by semiclassical models, in which light is still described by Maxwell's equations, but the material objects that emit and absorb light are quantized. Although these semiclassical models contributed to the development of quantum mechanics, further experiments proved Einstein's hypothesis that light itself is quantized; the quanta of light are photons.

In the modern *Standard Model of particle physics*, photons are described as a necessary consequence of physical laws having a certain symmetry at every point in spacetime. The intrinsic properties of photons, such as charge, mass and spin, are determined by the properties of this gauge symmetry.

The photon concept has led to momentous advances in experimental and theoretical physics, such as lasers, *Bose-Einstein condensation, quantum field theory*, and the probabilistic interpretation of quantum mechanics. It has been applied to photochemistry, high-resolution microscopy, and measurements of molecular distances. Recently, photons have been studied as elements of quantum computers and for sophisticated applications in optical communication such as quantum cryptography.

Nomenclature

In 1900 Max Planck was working on black-body radiation and suggested that the energy in electromagnetic waves could only be released in "packets" of energy, he called these quanta (singular quantum). Later, in 1905 Albert Einstein went further by suggesting that EM waves could only exist in these discrete wave-packets. He called these wave-packets photons. The name photon derives from the Greek word for light, fυ υ (transliterated phôs), and was coined in 1926 by the physical chemist Gilbert N. Lewis, who published a speculative theory in which photons were "uncreatable and indestructible". Although Lewis' theory was never accepted-being contradicted by many experiments-his new name, photon, was adopted immediately by most physicists. Isaac Asimov credits Arthur Compton with defining quanta of energy as photons in 1927.

In physics, a photon is usually denoted by the symbol? (the Greek letter gamma). This symbol for the photon probably derives from gamma rays, which were discovered and named in 1900 by Villard and shown to be a form of electromagnetic radiation in 1914 by Ernest Rutherford and Edward Andrade. In chemistry and optical engineering, photons are usually symbolized by hυ, the energy of a photon, where h is Planck's constant and the Greek letter υ (nu) is the photon's frequency. Much less commonly, the photon can be symbolized by hf, where its frequency is denoted by f.

Physical Properties

The photon is massless has no electric charge, and does not decay spontaneously in empty space. A photon has two possible polarization states and is described by exactly three continuous parameters: the components of its wave vector, which determine its wavelength ? and its direction of propagation. The photon is the gauge boson for electromagnetism and therefore all other quantum numbers- such as lepton number, baryon number, and flavour quantum numbers are zero.

Photons are emitted in many natural processes. For example, when a charge is accelerated it emits synchrotron radiation. During a molecular, atomic or nuclear transition to a lower energy level, photons of various energy will be emitted, from infrared light to gamma rays. A photon can also be emitted when a particle and its corresponding antiparticle are annihilated.

In empty space, the photon moves at c (the speed of light) and its energy and momentum are related by E = pc, where p is the magnitude of the momentum vector **p**. For comparison, the corresponding equation for particles with a mass m is

$$E^2 = p^2 c^2 + m^2 c^4$$

The energy and momentum of a photon depend only on its frequency (υ) or equivalently, its wavelength (λ):

$$E = \hbar\omega = h\nu = \frac{hc}{\lambda}$$

$$\mathbf{p} = \hbar\mathbf{k}$$

where **k** is the wave vector (with the wave number $k = |k| = 2\pi/\lambda$, $w = 2\pi/\lambda$ is the angular frequency, and $\hbar = h/2\pi$ is Dirac's constant or Planck's reduced constant.

Notice that **p** points in the direction of the photon's propagation. Consequently, the magnitude of the momentum is

$$p = \hbar k = \frac{h\nu}{c} = \frac{h}{\lambda}$$

The photon also carries spin angular momentum that does not depend on its frequency. The magnitude of its spin $\sqrt{2}\hbar$ is and the component measured along its direction of motion, its helicity, must be $\pm\hbar$. These two possible helicities correspond to the two possible circular polarization states of the photon (right-handed and left-handed).

To illustrate the significance of these formulae, the annihilation of a particle with its antiparticle must result in the creation of at least two photons for the following reason. In the center of mass frame, the colliding antiparticles have no net momentum, where as a single photon always has momentum (since it is determined, as we have seen, only by the photon's frequency or wavelength - which cannot be zero). Hence, conservation of momentum requires that at least two photons are created, with zero net momentum. The energy of the two photons—or, equivalently, their frequency—may be determined from conservation of four-momentum. Seen another way, the photon can be considered as its own antiparticle. The reverse process, pair production, is the dominant mechanism by which high-energy photons such as gamma rays lose energy while passing through matter.

The classical formulae for the energy and momentum of electromagnetic radiation can be re-expressed in terms of photon events. For example, the pressure of electromagnetic radiation on an object derives from the transfer of photon momentum per unit time and unit area to that object, since pressure is force per unit area and force is the change in momentum per unit time.

Historical Development

In most theories up to the eighteenth century, light was pictured as being made up of particles. One of the earliest particle theories was described in the Book of Optics (1021) by Alhazen, who held light rays to be streams of minute particles that "lack all sensible qualities except energy." Since particle models cannot easily account for the refraction, diffraction and birefringence of light, wave theories of light were proposed by René Descartes (1637), Robert Hooke (1665) nd Christian Huygens (1678) however, particle models remained dominant, chiefly due to the influence of Isaac Newton In the early nineteenth century, Thomas Young and August Fresnel clearly demonstrated the interference and diffraction of light and by 1850 wave models were generally accepted In 1865, James Clerk Maxwell's prediction that light was an electromagnetic wave—which was confirmed experimentally in 1888 by Heinrich Hertz's detection of radio waves—seemed to be the final blow to particle models of light.

The Maxwell wave theory, however, does not account for all properties of light. The Maxwell theory predicts that the energy of a light wave depends only on its intensity, not on its frequency; nevertheless, several independent types of experiments show that the energy imparted by light to atoms depends only on the light's frequency, not on its intensity. For example, some chemical reactions are provoked only by light of frequency higher than a certain threshold; light of frequency lower than the threshold, no matter how intense, does not initiate the reaction. Similarly, electrons can be ejected from a metal plate by shining light of sufficiently high frequency on it (the photoelectric effect); the energy of the ejected electron is related only to the light's frequency, not to its intensity.

At the same time, investigations of blackbody radiation carried culminated in Max Planck's hypothesis that the energy of any system that absorbs or emits electromagnetic radiation of frequency υ is an integer multiple of an energy quantum $E = h\upsilon$. As shown by Albert Einstein some form of energy quantization must be assumed to account for the thermal equilibrium observed between matter and electromagnetic radiation; for this explanation of the photoelectric effect, Einstein received the 1921 Nobel Prize in physics.

Since the Maxwell theory of light allows for all possible energies of electromagnetic radiation, most physicists assumed initially that the energy quantization resulted from some unknown constraint on the matter that absorbs or emits the radiation. In 1905, Einstein was the first to propose that energy quantization was a property of electromagnetic radiation itself. Although he accepted the validity of Maxwell's theory, Einstein pointed out that many anomalous experiments could be explained if the energy of a Maxwellian light wave were localized into point-like quanta that move independently of one another, even if the wave itself is spread continuously over space In 1909 and 1916, Einstein showed that, if Planck's law of black-body radiation is accepted, the energy quanta must also carry momentum $p = h/\upsilon$, making them full-fledged particles. This photon momentum was observed experimentally by Arthur Compton, for which he received the Nobel Prize in 1927. The pivotal question was then: how to unify Maxwell's wave theory of light with its experimentally observed particle nature? The answer to this question occupied Albert Einstein for the rest of his life and was solved in quantum electrodynamics and its successor, the Standard Model.

Early Objections

Einstein's 1905 predictions were verified experimentally in several ways in the first two decades of the 20th century, as recounted in Robert Millikan's Nobel lecture. however, before Compton's experiment showing that photons carried momentum proportional to their frequency (1922), most physicists were reluctant to believe that electromagnetic radiation itself might be particulate. for example, the Nobel lectures of Wien Planck and Millikan Instead, there was a widespread belief that energy quantization resulted from some unknown constraint on the matter that absorbs or emits radiation. Attitudes changed over gradually. In part, the change can be traced to experiments such as Compton scattering, where it was much more difficult not to ascribe quantization to light itself to explain the observed results.

Even after Compton's experiment, Bohr, Hendrik Kramers and John Slater made one last attempt to preserve the Maxwellian continuous electromagnetic field model of light, the so-called BKS model To account for the then-available data, two drastic hypotheses had to be made:

1. Energy and momentum are conserved only on the average in interactions between matter and radiation, not in elementary processes such as absorption and emission. This allows one to reconcile the discontinuously changing energy of the atom (jump between energy states) with the continuous release of energy into radiation.

2. Causality is abandoned. For example, spontaneous emissions are merely emissions induced by a "virtual" electromagnetic field.

However, refined Compton experiments showed that energy-momentum is conserved extraordinarily well in elementary processes; and also that the jolting of the electron and the generation of a new photon in Compton scattering obey causality to within 10 ps (pico second). Accordingly, Bohr and his co-workers gave their model "as honorable a funeral as possible". Nevertheless, the failures of the BKS model inspired Werner Heisenberg in his development of matrix mechanics.

A few physicists persisted in developing semiclassical models in which electromagnetic radiation is not quantized, but matter obeys the laws of quantum mechanics. Although the evidence for photons from chemical and physical experiments was overwhelming by the 1970s, this evidence could not be considered as absolutely definitive; since it relied on the interaction of light with matter, a sufficiently complicated theory of matter could in principle account for the evidence. Nevertheless, all semiclassical theories were refuted definitively in the 1970s and 1980s by photon-correlation experiments Hence, Einstein's hypothesis that quantization is a property of light itself is considered to be proven.

Wave-particle Duality and Uncertainty Principles

Photons, like all quantum objects, exhibit both wave-like and particle-like properties. Their dual wave–particle nature can be difficult to visualize. The photon displays clearly wave-like phenomena such as diffraction and interference on the length scale of its wavelength. For example, a single photon passing through a double-slit experiment lands on the screen with a probability distribution given by its interference pattern determined by Maxwell's equations However, experiments confirm that the photon is not a short pulse of electromagnetic radiation; it does not spread out as it propagates, nor does it divide when it encounters a beam splitter. Rather, the photon seems to be a point-like particle since it is absorbed or emitted as a whole by arbitrarily small systems, systems much smaller than its wavelength, such as an atomic nucleus (≈ 10-15 m across) or even the point-like electron. Nevertheless, the photon is not a point-like particle whose trajectory is shaped probabilistically by the electromagnetic field, as conceived by Einstein and others; that hypothesis was also refuted by the photon-correlation experiments cited above. According to our present understanding, the electromagnetic field itself is produced by photons, which in turn result from a local gauge symmetry and the laws of quantum field theory.

A key element of quantum mechanics is Heisenberg's uncertainty principle, which forbids the simultaneous measurement of the position and momentum of a particle along the same direction. Remarkably, the uncertainty principle for charged, material particles requires the quantization of light into photons, and even the frequency dependence of the photon's energy and momentum. An elegant illustration is Heisenberg's thought experiment for locating an electron with an ideal microscope. The position of the electron can be determined to within the resolving power of the microscope, which is given by a formula from classical optics.

$$\Delta x \sim \frac{\lambda}{\sin\theta}$$

where θ is the aperture angle of the microscope. Thus, the position uncertainty Δx can be made arbitrarily small by reducing the wavelength λ. The momentum of the electron is uncertain, since it received a "kick" Δp from the light scattering from it into the microscope. If light were not quantized into photons, the uncertainty Δp could be made arbitrarily small by reducing the light's intensity. In that case, since the wavelength and intensity of light can be varied independently, one could simultaneously determine the position and momentum to arbitrarily high accuracy, violating the uncertainty principle. By contrast, Einstein's formula for photon momentum preserves the uncertainty principle; since the photon is scattered anywhere within the aperture, the uncertainty of momentum transferred equals

$$\Delta p \sim p_{\text{photon}} \sin\theta = \frac{h}{\lambda} \sin\theta$$

giving the product, $\Delta x \Delta p \sim h$ which is Heisenberg's uncertainty principle. Thus, the entire world is quantized; both matter and fields must obey a consistent set of quantum laws, if either one is to be quantized.

The analogous uncertainty principle for photons forbids the simultaneous measurement of the number n of photons in an electromagnetic wave and the phase φ of that wave.

Both photons and material particles such as electrons create analogous interference patterns when passing through a double-slit experiment. For photons, this corresponds to the interference of a Maxwell light wave whereas, for material particles, this corresponds to the interference of the Schrödinger wave equation. Although this similarity might suggest that Maxwell's equations are simply Schrödinger's equation for photons, most physicists do not agree. For one thing, they are mathematically different; most obviously, Schrödinger's one equation solves for a complex field, whereas Maxwell's four equations solve for real fields. More generally, the normal concept of a Schrödinger probability wave function cannot be applied to photons. Being massless, they cannot be localized without being destroyed; technically, photons cannot have a position eigenstate $|r\rangle$, and, thus, the normal Heisenberg uncertainty principle $\Delta x \Delta p > h/2$ does not pertain to photons. A few substitute wave functions have been suggested for the photon, but they have not come into general use. Instead, physicists generally accept the second-quantized theory of photons described below, quantum electrodynamics, in which photons are quantized excitations of electromagnetic modes.

Bose-Einstein Model of a Photon Gas

In 1924, Satyendra Nath Bose derived Planck's law of black-body radiation without using any electromagnetism, but rather a modification of coarse-grained counting of phase space Einstein showed that this modification is equivalent to assuming that photons are rigorously identical and that it implied a "mysterious non-local interaction", now understood as the requirement for a symmetric quantum mechanical state. This work led to the concept of coherent states and the development of the laser. In the same papers, Einstein extended Bose's formalism to material particles (bosons) and predicted that they would condense into their lowest quantum state at low enough temperatures; this Bose-Einstein condensation was observed experimentally in 1995.

The modern view on this is that photons are, by virtue of their integer spin, bosons (as opposed to fermions with half-integer spin). By the spin-statistics theorem, all bosons obey Bose–Einstein statistics (whereas all fermions obey Fermi-Dirac statistics).

Stimulated and Spontaneous Emission

In 1916, Einstein showed that Planck's radiation law could be derived from a semi-classical, statistical treatment of photons and atoms, which implies a relation between the rates at which atoms emit and absorb photons. The condition follows from the assumption that light is emitted and absorbed by atoms independently, and that the thermal equilibrium is preserved by interaction with atoms. Consider a cavity in thermal equilibrium and filled with electromagnetic radiation and atoms that can emit and absorb that radiation. Thermal equilibrium requires that the energy density ρ(ν) of photons with frequency ν (which is proportional to their number density) is, on average, constant in time; hence, the rate at which photons of any particular frequency are emitted must equal the rate of absorbing them.

Einstein began by postulating simple proportionality relations for the different reaction rates involved. In his model, the rate Rji for a system to absorb a photon of frequency ν and transition from a lower energy E_j to a higher energy E_i is proportional to the number N_j of atoms with energy E_j and to the energy density ρ(ν) of ambient photons with that frequency,

$$R_{ji} = N_j\, B_{ji} p(v)$$

where B_{ji} is the rate constant for absorption. For the reverse process, there are two possibilities: spontaneous emission of a photon, and a return to the lower-energy state that is initiated by the interaction with a passing photon. Following Einstein's approach, the corresponding rate Rij for the emission of photons of frequency ν and transition from a higher energy E_i to a lower energy E_j is

$$R_{ij} = N_i A_{ij} + N_i B_{ij} p(v)$$

where A_{ij} is the rate constant for emitting a photon spontaneously, and Bij is the rate constant for emitting it in response to ambient photons (induced or stimulated emission). In thermodynamic equilibrium, the number of atoms in state i and that of atoms in state j must, on average, be constant; hence, the rates R_{ji} and R_{ij} must be equal. Also, by arguments analogous to the derivation of Boltzmann statistics, the ratio of N_i and N_j is $g_i/g_j \exp(E_j - E_i) / kT)$, where $g_{i,j}$ are the degeneracy of the state i and that of j, respectively, $E_{i,j}$ their energies, k the Boltzmann constant and T the system's temperature. From this, it is readily derived that $g_i B_{ij} = g_j B_{ji}$ and

$$A_{ij} = \frac{8\pi h v^3}{c^3} B_{ij}$$

The A and Bs are collectively known as the Einstein coefficients.

Einstein could not fully justify his rate equations, but claimed that it should be possible to calculate the coefficients A_{ij}, B_{ji} and B_{ij} once physicists had obtained "mechanics and electrodynamics modified to accommodate the quantum hypothesis" In fact, in 1926, Paul Dirac derived the B_{ij} rate constants in using a semiclassical approach, and, in 1927, succeeded in deriving all the rate constants from first principles within the framework of quantum theory Dirac's work was the foundation of quantum electrodynamics, i.e., the quantization of the electromagnetic field itself. Dirac's approach is also called *second quantization* or *quantum field theory* earlier quantum mechanical treatments only treat material particles as quantum mechanical, not the electromagnetic field.

Einstein was troubled by the fact that his theory seemed incomplete, since it did not determine the direction of a spontaneously emitted photon. A probabilistic nature of light-particle motion was first considered by Newton in his treatment of birefringence and, more generally, of the splitting of light beams at interfaces into a transmitted beam and a reflected beam. Newton hypothesized that hidden variables in the light particle determined which path it would follow. Similarly, Einstein hoped for a more complete theory that would leave nothing to chance, beginning his separation from quantum mechanics. Ironically, Max Born's probabilistic interpretation of the wave function was inspired by Einstein's later work searching for a more complete theory.

Second Quantization

In 1910, Peter Debye derived Planck's law of black-body radiation from a relatively simple assumption. He correctly decomposed the electromagnetic field in a cavity into its Fourier modes, and assumed that the energy in any mode was an integer multiple of $h\upsilon$, where υ is the frequency of the electromagnetic mode. Planck's law of black-body radiation follows immediately as a geometric sum. However, Debye's approach failed to give the correct formula for the energy fluctuations of blackbody radiation, which were derived by Einstein in 1909.

In 1925, Born, Heisenberg and Jordan reinterpreted Debye's concept in a key way. As may be shown classically, the Fourier modes of the electromagnetic field—a complete set of electromagnetic plane waves indexed by their wave vector **k** and polarization state—are equivalent to a set of uncoupled simple harmonic oscillators. Treated quantum mechanically, the energy levels of such oscillators are known to be $E = nh\upsilon$, where υ is the oscillator frequency. The key new step was to identify an electromagnetic mode with energy $E = nh\upsilon$ as a state with n photons, each of energy $h\upsilon$. This approach gives the correct energy fluctuation formula.

Dirac took this one step further. He treated the interaction between a charge and an electromagnetic field as a small perturbation that induces transitions in the photon states, changing the numbers of photons in the modes, while conserving energy and momentum overall. Dirac was able to derive Einstein's A_{ij} and B_{ij} coefficients from first principles, and showed that the Bose–Einstein statistics of photons is a natural consequence of quantizing the electromagnetic field correctly (Bose's reasoning went in the opposite direction; he derived Planck's law of black body radiation by assuming BE statistics). In Dirac's time, it was not yet known that all bosons, including photons, must obey BE statistics.

Dirac's second-order perturbation theory can involve virtual photons, transient intermediate states of the electromagnetic field; the static electric and magnetic interactions are mediated by such virtual photons. In such quantum field theories, the probability amplitude of observable events is calculated by summing over all possible intermediate steps, even ones that are unphysical; hence, virtual photons are not constrained to satisfy E = pc, and may have extra polarization states; depending on the gauge used, virtual photons may have three or four polarization states, instead of the two states of real photons. Although these transient virtual photons can never be observed, they contribute measurably to the probabilities of observable events. Indeed, such second-order and higher-order perturbation calculations can give apparently infinite contributions to the sum. Such unphysical results are corrected for using the technique of renormalization. Other virtual particles may contribute to the summation as well; for example, two photons may interact indirectly through virtual electron-positron pairs. In fact, such photon-photon scattering, as well as electron-photon scattering, is meant to be one of the modes of operations of the planned particle accelerator, the International Linear Collider.

In modern physics notation, the quantum state of the electromagnetic field is written as a Fock state, a tensor product of the states for each electromagnetic mode

$$|n_{ko}\rangle \otimes |n_{k1}\rangle \otimes \ldots \otimes |n_{kn}\rangle \ldots$$

where $|n_{ki}\rangle$ represents the state in which $|n_{ki}\rangle$, photons are in the mode ki. In this notation, the creation of a new photon inmode ki (e.g., emitted from an atomic transition) is written as $|n_{ki}\rangle \rightarrow |n_{ki}{+}1\rangle$. This notation merely express the concept of Born, Heisenberg and Jordan described above, and does not add any physics.

The Photon as a Gauge Boson

The electromagnetic field can be understood as a gauge theory, i.e., as a field that results from requiring that symmetry hold independently at every position in spacetime. For the electromagnetic field, this gauge symmetry is the Abelian U(1) symmetry of a complex number, which reflects the ability to vary the phase of a complex number without affecting real numbers made from it, such as the energy or the Lagrangian.

The quanta of an Abelian gauge field must be massless, uncharged bosons, as long as the symmetry is not broken; hence, the photon is predicted to be massless, and to have zero electric charge and integer spin. The particular form of the electromagnetic interaction specifies that the photon must have spin ±1; thus, its helicity must be $\pm \hbar$. These two spin components correspond to the classical concepts of right-handed and left-handed circularly polarized light. However, the transient virtual photons of quantum electrodynamics may also adopt unphysical polarization states.

In the prevailing Standard Model of physics, the photon is one of four gauge bosons in the electroweak interaction; the other three are denoted W^+, W^- and Z^0 and are responsible for the weak interaction. Unlike the photon, these gauge bosons have invariant mass, owing to a mechanism that breaks their SU(2) gauge symmetry. The unification of the photon with W and Z gauge bosons in the electroweak interaction was accomplished by Sheldon Glashow, Abdus Salam and Steven Weinberg, for which they were awarded the 1979 Nobel Prize in physics Physicists continue to hypothesize grand unified theories that connect these four gauge bosons with the eight gluon gauge bosons of quantum chromodynamics; however, key predictions of these theories, such as proton decay, have not been observed experimentally.

Contributions to the Mass of a System

The energy of a system that emits a photon is decreased by the energy E of the photon as measured in the rest frame of the emitting system, which may result in a reduction in mass in the amount E/c^2. Similarly, the mass of a system that absorbs a photon is increased by a corresponding amount. As an application, the energy balance of nuclear reactions involving photons is commonly written in terms of the masses of the nuclei involved, and terms of the form E/c^2 for the gamma photons (and for other relevant energies, such as the recoil energy of nuclei).

This concept is applied in key predictions of quantum electrodynamics (QED, see above). In that theory, the mass of electrons (or, more generally, leptons) is modified by including the mass contributions of virtual photons, in a technique known as renormalization. Such "radiative corrections" contribute to a number of predictions of QED, such as the magnetic dipole moment of leptons, the Lamb shift, and the hyperfine structure of bound lepton pairs, such as muonium and positronium.

Since photons contribute to the stress-energy tensor, they exert a gravitational attraction on other objects, according to the theory of general relativity. Conversely, photons are themselves affected by gravity; their normally straight trajectories may be bent by warped spacetime, as in gravitational lensing, and their frequencies may be lowered by moving to a higher gravitational potential, as in the Pound-Rebka experiment. However, these effects are not specific to photons; exactly the same effects would be predicted for classical electromagnetic waves.

Photons in Matter

Light that travels through transparent matter does so at a lower speed than c, the speed of light in a vacuum. For example, photons suffer so many collisions on the way from the core of the sun that radiant energy can take about a million years to reach the surface however, once in open space, a photon takes only 8.3 minutes to reach Earth. The factor by which the speed of light is decreased is called the refractive index of the material. In a classical wave picture, the slowing can be explained by the light inducing electric polarization in the matter, the polarized matter radiating new light, and the new light interfering with the original light wave to form a delayed wave. In a particle picture, the slowing can instead be described as a blending of the photon with quantum excitations of the matter (quasi-particles such as phonons and excitons) to form a polariton; this polariton has a nonzero effective mass, which means that it cannot travel at c.

Alternatively, photons may be viewed as always traveling at c, even in matter, but they have their phase shifted (delayed or advanced) upon interaction with atomic scatters: this modifies their wavelength and momentum, but not speed. A light wave made up these photons does travel slower than the speed of light. In this view the photons are "bare", and are scattered and phase shifted, while in the view of the preceding paragraph the photons are "dressed" by their interaction with matter, and move without scattering or phase shifting, but at a lower speed.

Light of different frequencies may travel through matter at different speeds; this is called dispersion. In some cases, it can result in extremely slow speeds of light in matter. The effects of photon interactions with other quasi-particles may be observed directly in Raman scattering and Brillouin scattering.

Photons can also be absorbed by nuclei, atoms or molecules, provoking transitions between their energy levels. A classic example is the molecular transition of retinal $C_{20}H_{28}O$, which is responsible for vision, as discovered in 1958 by Nobel laureate biochemist George Wald and co-workers. As shown here, the absorption provokes a cis-trans isomerization that, in combination with other such transitions, is transduced into nerve impulses. The absorption of photons can even break chemical bonds, as in the photodissociation of chlorine; this is the subject of photochemistry.

Technological Applications

Photons have many applications in technology. These examples are chosen to illustrate applications of photons *per se*, rather than general optical devices such as lenses, etc. that could operate under a classical theory of light. The laser is an extremely important application and is discussed above under stimulated emission.

Individual photons can be detected by several methods. The classic photomultiplier tube exploits the photoelectric effect: a photon landing on a metal plate ejects an electron, initiating an ever-amplifying avalanche of electrons. Charge-coupled device chips use a similar effect in semiconductors: an incident photon generates a charge on a microscopic capacitor that can be detected. Other detectors such as Geiger counters use the ability of photons to ionize gas molecules, causing a detectable change in conductivity.

Planck's energy formula $E = h\upsilon$ is often used by engineers and chemists in design, both to compute the change in energy resulting from a photon absorption and to predict the frequency of the light emitted for a given energy transition. For example, the emission spectrum of a fluorescent light bulb can be designed using gas molecules with different electronic energy levels and adjusting the typical energy with which an electron hits the gas molecules within the bulb.

Under some conditions, an energy transition can be excited by two photons that individually would be insufficient. This allows for higher resolution microscopy, because the sample absorbs energy only in the region where two beams of different colors overlap significantly, which can be made much smaller than the excitation volume of a single beam (see two-photon excitation microscopy). Moreover, these photons cause less damage to the sample, since they are of lower energy.

In some cases, two energy transitions can be coupled so that, as one system absorbs a photon, another nearby system "steals" its energy and re-emits a photon of a different frequency. This is the basis of fluorescence resonance energy transfer, a technique that is used in molecular biology to study the interaction of suitable proteins.

Several different kinds of hardware random number generator involve the detection of single photons. In one example, for each bit in the random sequence that is to be produced, a photon is sent to a beam-splitter. In such a situation, there are two possible outcomes of equal probability. The actual outcome is used to determine whether the next bit in the sequence is "0" or "1"

Recent Research

Much research has been devoted to applications of photons in the field of quantum optics. Photons seem well-suited to be elements of an ultra-fast quantum computer, and the quantum entanglement of photons is a focus of research. Nonlinear optical processes are another active research area, with topics such as two-photon absorption, self-phase modulation and optical parametric oscillators. However, such processes generally do not require the assumption of photons per se; they may often be modeled by treating atoms as nonlinear oscillators. The nonlinear process of spontaneous parametric down conversion is often used to produce single-photon states. Finally, photons are essential in some aspects of optical communication, especially for quantum cryptography.

DOPPLER EFFECT

The Doppler effect (or Doppler shift), named after Austrian physicist Christian Doppler who proposed it in 1842, is the change in frequency and wavelength of a wave for an observer moving relative to the source of the waves. It is commonly heard when a vehicle sounding a siren approaches, passes and recedes from an observer. The received frequency is increased (compared to the emitted frequency) during the approach, it is identical at the instant of passing by, and it is decreased during the recession.

For waves that propagate in a medium, such as sound waves, the velocity of the observer and of the source are relative to the medium in which the waves are transmitted. The total Doppler effect may therefore result from motion of the source, motion of the observer, or motion of the medium. Each of these effects is analyzed separately. For waves which do not require a medium, such as light or gravity in special relativity, only the relative difference in velocity between the observer and the source needs to be considered.

Development

Doppler first proposed the effect in 1842 in his treatise "Über das farbige Licht der Doppelsterne und einiger anderer Gestirne des Himmels" (On the coloured light of the binary stars and some other stars of the heavens) The hypothesis was tested for sound waves by Buys Ballot in 1845. He confirmed that the sound's pitch was higher than the emitted frequency when the sound source approached him, and lower than the emitted frequency when the sound source receded from him. Hippolyte Fizeau discovered independently the same phenomenon on electromagnetic waves in 1848 (in France, the effect is sometimes called "effet Doppler-Fizeau"). In Britain, John Scott Russell made an experimental study of the Doppler effect (1848). An English translation of Doppler's 1842 treatise can be found in the book *The Search for Christian Doppler* by *Alec Eden*.

General

The relationship between observed frequency f and emitted frequency f_0 is given by

$$f = \left(\frac{v + v_r}{v + v_r}\right) f_0$$

where

v is the velocity of waves in the medium

v is the velocity of the receive relative to the medium; positive if the receiver is moving towards the source.

$v_{s,r}$ is the velocity of the source relative to the medium, positive if the source is moving away from the receiver.

The observed frequency is increased if the source is moving towards the observer. It is decreased if the source is moving away.

For all paths of an approaching object, the observed frequency that is first heard is higher than the object's emitted frequency.

Thereafter there is a monotonic decrease in the observed frequency as it gets closer to the observer, through equality when it is level with the observer, and a continued monotonic decrease as it recedes from the observer. When the observer is very close to the path of the object, the transition from high to low frequency is very abrupt. When the observer is far from the path of the object, the transition from high to low frequency is gradual.

In the limit where the speed of the wave is much greater than the relative speed of the source and observer (this is often the case with electromagnetic waves, e.g. light), the relationship between observed frequency f; and emitted frequency f0 is given by:

Change in Frequency

$$\Delta f = \frac{f_0 v_{s,r}}{c} = -\frac{v_{s,r}}{\lambda_0}$$

Observed Frequency

$$f = \left(1 - \frac{v_{s,r}}{c}\right) f_0$$

where

$v_{s,r}$ is the velocity of the transmitter relative to the receiver in meters per second: negative when moving towards one another, positive when moving away.

c is the speed of wave (e.g. 3×10^8 m/s for electromagnetic waves travelling in a vacuum)

λ_0 is the wavelength of the transmitted wave subject to change.

These two equations are only accurate to a first order approximation. However, they work reasonably well in the case considered by Doppler: when the speed between the source and receiver is slow relative to the speed of the waves involved and the distance between the source and receiver is large relative to the wavelength of the waves. If either of these two approximations are violated, the formulae are no longer accurate.

Analysis

The frequency of the sounds that the source emits does not actually change. To understand what happens, consider the following analogy. Someone throws one ball every second in a man's direction. Assume that balls travel with constant velocity. If the thrower is stationary, the man will receive one ball every second. However, if the thrower is moving towards the man, he will receive balls more frequently because the balls will be less spaced out. The inverse is true if the thrower is moving away from the man. So it is actually the wavelength which is affected; as a consequence, the received frequency is also affected. It may also be said that the velocity of the wave remains constant whereas wavelength changes; hence frequency also changes.

If the moving source is emitting waves through a medium with an actual frequency f0, then an observer stationary relative to the medium detects waves with a frequency f given by

$$f = f_0\left(\frac{v}{v + v_s}\right)$$

where υ_s is the speed of the waves in the medium and vs is the speed of the source with respect to the medium (positive if moving away from the observer, negative if moving towards the observer).

A similar analysis for a moving observer and a stationary source yields the observed frequency (the receiver's velocity being represented as υ_r):

$$f = f_0\left(\frac{v}{v + v_r}\right)$$

where the same convention applies: υ_r is positive if the observer is moving away from the source, and negative if the observer is moving towards the source.

These can be generalized into a single equation with both the source and receiver moving.

$$f = \left(\frac{v}{v + v_{s,r}}\right)f_0$$

which can be written as:

$$f = \left(1 - \frac{v_{s,r}}{v + v_{s,r}}\right)f_0$$

where $\upsilon_{s,r}$ is the source to receiver velocity radial component.

With a relatively slow moving source, vs,r is small in comparison to v and the equation approximates to

$$f = \left(1 - \frac{v_{s,r}}{v}\right) f_0$$

However the limitations mentioned above still apply. When the more complicated exact equation is derived without using any approximations (just assuming that everything: source, receiver, and wave or signal are moving linearly) several interesting and perhaps surprising results are found. For example, as Lord Rayleigh noted in his classic book on sound, by properly moving it would be possible to hear a symphony being played backwards. This is the so-called "time reversal effect" of the Doppler effect. Other interesting cases are that the Doppler effect is time dependent in general (thus we need to know not only the source and receivers' velocities, but also their positions at a given time) and also in some circumstances it is possible to receive two signals or waves from a source (or no signal at all). In addition there are more possibilities than just the receiver approaching the signal and the receiver receding from the signal.

All these additional complications are for the classical—i.e., non-relativistic Doppler effect. However, all these results also hold for the relativistic Doppler effect as well.

The first attempt to extend Doppler's analysis to light waves was soon made by Fizeau. In fact, light waves do not require a medium to propagate and the correct understanding of the Doppler effect for light requires the use of the Special Theory of Relativity.

A Common Misconception

Craig Bohren pointed out that some physics textbooks erroneously state that the observed frequency increases as the object approaches an observer and then decreases only as the object passes the observer. In fact, the observed frequency of an approaching object declines monotonically from a value above the emitted frequency, through a value equal to the emitted frequency when the object is closest to the observer, and to values increasingly below the emitted frequency as the object recedes from the observer. Bohren proposed that this common misconception might occur because the intensity of the sound increases as an object approaches an observer and decreases once it passes and recedes from the observer.

Applications

Sirens

The siren on a passing emergency vehicle will start out higher than its stationary pitch, slide down as it passes, and continue lower than its stationary pitch as it recedes from the observer. Astronomer John Dobson explained the effect thus:

> "The reason the siren slides is because it doesn't hit you."

In other words, if the siren approached the observer directly, the pitch would remain constant (as vs, r is only the radial component) until the vehicle hit him, and then immediately jump to a new lower pitch. Because the vehicle passes by the observer, the radial velocity does not remain constant, but instead varies as a function of the angle between his line of sight and the siren's velocity:

$$v_{s,r} = v_s . \cos\theta$$

where υ_s is the velocity of the object (source of waves) with respect to the medium, and θ is the angle between the object's forward velocity and the line of sight from the object to the observer.

Astronomy

The Doppler effect for electromagnetic waves such as light is of great use in astronomy and results in either a so-called redshift or blueshift. It has been used to measure the speed at which stars and galaxies are approaching or receding from us, that is, the radial velocity. This is used to detect if an apparently single star is, in reality, a close binary and even to measure the rotational speed of stars and galaxies.

The use of the Doppler effect for light in astronomy depends on our knowledge that the spectra of stars are not continuous. They exhibit absorption lines at well defined frequencies that are correlated with the energies required to excite electrons in various elements from one level to another. The Doppler effect is recognizable in the fact that the absorption lines are not always at the frequencies that are obtained from the spectrum of a stationary light source. Since blue light has a higher frequency than red light, the spectral lines of an approaching astronomical light source exhibit a blueshift and those of a receding astronomical light source exhibit a redshift.

Among the nearby stars, the largest radial velocities with respect to the Sun are +308 km/s (BD-15° 4041, also known as LHS 52, 81.7 light-years away) and -260 km/s (Woolley 9722, also known as Wolf 1106 and LHS 64, 78.2 light-years away). Positive radial velocity means the star is receding from the Sun, negative that it is approaching.

Temperature Measurement

Another use of the Doppler effect, which is found mostly in astronomy, is the estimation of the temperature of a gas which is emitting a spectral line. Due to the thermal motion of the gas, each emitter can be slightly red or blue shifted, and the net effect is a broadening of the line. This line shape is called a Doppler profile and the width of the line is proportional to the square root of the temperature of the gas, allowing the Doppler-broadened line to be used to measure the temperature of the emitting gas.

Radar

The Doppler effect is used in some types of radar, to measure the velocity of detected objects. A radar beam is fired at a moving target—e.g. a motor car, as police use radar to detect speeding motorists—as it approaches or recedes from the radar source. Each successive radar wave has to travel farther to reach the car, before being reflected and re-detected near the sou wave increases, increasing the wavelength. In some situations, the radar beam is fired at the moving car as it approaches, in which case each successive wave travels a lesser dista II, relies upon Doppler radar to explode at the correct time, height, distance, et cetera.

Medical Imaging and Blood Flow Measurement

An echocardiogram can, within certain limits, produce accurate assessment of the direction of blood flow and the velocity of bar regurgitation), and calculation of the cardiac output. Contrast-enhanced ultrasound using gas-filled microbubble contrast media can be used to improve velocity or other flow-related medical measurements.

Although "Doppler" has become synonymous with velocity measurement in medical imaging in many cases it is not the frequency shift (Doppler shift) of the received signal that is measured, but the phase shift (when the received signal arrives).

Velocity measurements of blood flow are also used in other fields of medical ultrasonography, such as obstetric ultrasonography and neurology. Velocity measurement of blood flow in arteries and veins based on Doppler effect is an effective tool for diagnosis of vascular problems like stenosis.

Flow Measurement

Instruments such as the laser Doppler velocimeter (LDV), and Acustic Doppler Velocimeter (ADV) have been developed to measure velocities in a fluid flow. The LDV emits a light beam and the ADV emits an ultrasonic acoustic burst, and measure water velocity and face. This technique allows non-intrusive flow measurements, at high precision and high frequency.

Velocity Profile Measurement

Developed originally for velocity measurements in medical applications (blood flows), Ultrasonic Doppler Velocimetry (UDV) can measure in real time complete velocity profile in almost any liquids containing particles in suspension such as dust, gas bubbles, emulsions. Flows can be pulsating, oscillating, laminar or turbulent, stationary or transient. This technique is fully non-invasive.

Underwater Acoustics

In military applications the Doppler shift of a target is used to ascertain the speed of a submarine using both passive and active sonar systems. As a submarine passes by a passive sonobuoy, the stable frequencies undergo a Doppler shift, and the speed and range from the sonobuoy can be calculated. If the sonar system is mounted on a moving ship or another submarine, then the relative velocity can be calculated.

Audio

The Leslie speaker, associated with and predominantly used with the Hammond B-3 Organ, takes advantage of the Doppler Effect by using an electric motor to rotate a speaker continuously, rapidly alternating the received frequency of a keyboard note.

RELATIVISTIC DOPPLER EFFECT

The relativistic Doppler effect is the change in frequency (and wavelength) of light, caused by the relative motion of the source and the observer (as in the classical Doppler effect), when taking into account effects of the special theory of relativity.

The relativistic Doppler effect is different from the non-relativistic Doppler effect as the equations include the time dilation effect of special relativity and do not involve the medium of propagation as a reference point. They describe the total difference in observed frequencies and possess the required Lorentz symmetry.

THE MECHANISM (A SIMPLE CASE)

Assume the observer and the source are moving away from each other with a relative velocity V (the sign of V is simply reversed in the case where the observers are moving toward each other). Let us consider the problem from the reference frame of the source.

Suppose one wavefront arrives at the observer. The next wavefront is then at a distance $\lambda = c/f_e$ away from him (where λ is the wavelength, f_e is the frequency of the wave the source emitted, and C is the speed of light). Since the wavefront moves with velocity and the observer escapes with velocity V, the time observed between crests is

$$t = \frac{\lambda}{c\text{-}v} = \frac{1}{\left(1\text{-}\frac{v}{c}\right)f_e}$$

However, due to the relativistic time dilation, the observer will measure this time to be

$$t_0 = \frac{t}{\gamma} = \frac{1}{\gamma\left(1\text{-}\frac{v}{c}\right)f_e}$$

where $\gamma = 1\Big/\sqrt{1 - \frac{v^2}{c^2}} = 1\Big/\sqrt{\left(1 - \frac{v}{c}\right)\left(1 + \frac{v}{c}\right)}$

so the corresponding observed frequency is

$$f_0 = \frac{1}{t_0} = \gamma\left(1 - \frac{v}{c}\right)f_e = \sqrt{\frac{\left(1 - \frac{v}{c}\right)}{\left(1 + \frac{v}{c}\right)}}f_e$$

The ratio f_e/f_o is called the Doppler factor of the source relative to the observer. (This terminology is particularly prevalent in the subject of astrophysics: see relativistic beaming.)

For Motion Along the Line of Sight

If the observer and the source are moving directly towards each other with velocity , the observed frequency f_o is different from the frequency of the source f_e as

$$f_0 = \sqrt{\frac{\left(1 + \frac{v}{c}\right)}{\left(1 - \frac{v}{c}\right)}}f_e$$

If the observer and the source are moving directly away from each other with velocity , , the observed frequency f_o is different from the frequency of the source f_e as

$$f_0 = \sqrt{\frac{\left(1 - \frac{v}{c}\right)}{\left(1 + \frac{v}{c}\right)}}f_e$$

where C is the speed of light.

The corresponding wavelengths are related by

$$\lambda_0 = \sqrt{\frac{\left(1 + \frac{v}{c}\right)}{\left(1 - \frac{v}{c}\right)}}\lambda_e$$

and the resulting redshift can be written as

$$z + 1 = \frac{\lambda_0}{\lambda_e} = \sqrt{\frac{(1 + v/c)}{(1 - v/c)}}$$

In the non-relativistic limit—i.e. when $v \ll c$—the approximate expressions are

$$\frac{\Delta f}{f} \approx -\frac{v}{c}; \frac{\Delta\lambda}{\lambda} \approx \frac{v}{c}; z \approx \frac{v}{c}$$

For Motion in an Arbitrary Direction

If, in the reference frame of the observer, the source is moving away with velocity V at an angle θ_0 relative to the direction from the observer to the source (at the time when the light is emitted), the frequency changes as

$$f_0 = \frac{f_s}{\gamma\left(1 + \frac{v\cos\theta_o}{c}\right)} \quad (2.1)$$

where $\gamma = \frac{1}{\sqrt{1 - v^2/c^2}}$

In the particular case when $\theta_o = 90°$ and $\cos\theta_o = 0$ one obtains the transverse Doppler effect

$$f_0 = \frac{f_s}{\gamma} \quad (2.2)$$

However, if the angle θ_s is measured in the reference frame of the source (at the time when the light is received by the observer), the expression is

$$f_0 = \gamma\left(1 - \frac{v\cos\theta_s}{c}\right)f_s \quad (2.3)$$

$\cos\theta_o$ and $\cos\theta_s$ are tied to each other via the relativistic aberration formula:

$$\cos\theta_0 = \frac{\cos\theta_s - \frac{v}{c}}{1 - \frac{v}{c}\cos\theta_s} \quad (2.4)$$

The relativistic aberration formula explains why, for one obtains a second formula for the transverse Doppler effect:

$$f_0 = f_s\gamma \quad (2.5)$$

(2.5) is obtained easily by substituting $\cos\theta_0 = -\frac{v}{c}$ into (2.1). Turns out that (2.5) is more useful than (2.2) being the form used routinely in the Ives-Stilwell experiment.

In the non-relativistic limit, both formulæ become

$$\frac{\Delta f}{f} \approx \frac{v\cos\theta}{c}$$

CHAPTER 3

Redshift

In physics and astronomy, redshift occurs when electromagnetic radiation—usually visible light—emitted or reflected by an object is shifted towards the (less energetic) red end of the electromagnetic spectrum due to the Doppler effect. More generally, redshift is defined as an increase in the wavelength of electromagnetic radiation received by a detector compared with the wavelength emitted by the source. This increase in wavelength corresponds to a drop in the frequency of the electromagnetic radiation. Conversely, a decrease in wavelength is called blue shift.

Any increase in wavelength is called "redshift", even if it occurs in electromagnetic radiation of non-optical wavelengths, such as gamma rays, x-rays and ultraviolet. This nomenclature might be confusing since, at wavelengths longer than red (e.g., infrared, microwaves, and radio waves), redshifts shift the radiation away from the red wavelengths.

An observed redshift due to the Doppler effect occurs whenever a light source moves away from the observer, corresponding to the Doppler shift that changes the perceived frequency of sound waves. Although observing such redshifts, or complementary blue shifts, has several terrestrial applications (e.g., Doppler radar and radar guns), spectroscopic astrophysics uses Doppler redshifts to determine the movement of distant astronomical objects.

The special relativistic redshift formula (and its Newtonian approximation) applies only when spacetime is flat. Where gravitational effects are important, redshift must be calculated using general relativity. Two important special-case formulas are the so-called gravitational redshift formula which applies to any stationary (that is, unchanging with time) gravitational field, and the cosmological redshift formula which applies to the expanding universe of Big Bang cosmology.

Special relativistic, gravitational, and cosmological redshift can be understood under the umbrella of frame transformation laws, as described below. There exist numerous other mechanisms with different physical and mathematical descriptions that can lead to a shift in the frequency of electromagnetic radiation and whose action is generally not referred to as a "redshift", including scattering and optical effects (for more see section on physical optics and radiative transfer).

History

The history of the subject began with the development in the 19th century of wave mechanics and the exploration of phenomena associated with the Doppler effect. The effect is named after Christian Andreas Doppler, who offered the first known physical explanation for the phenomenon in 1842. The hypothesis was tested and confirmed for sound waves by the Dutch scientist Christoph Hendrik Diederik Buys Ballot in 1845 Doppler correctly predicted that the phenomenon should apply to all waves, and in particular suggested that the varying colors of stars could be attributed to their motion with respect to the Earth While this attribution turned out to be incorrect (stellar colors are indicators of a star's temperature, not motion), Doppler would later be vindicated by verified redshift observations.

The first Doppler redshift was described in 1848 by French physicist Armand-Hippolyte-Louis Fizeau, who pointed to the shift in spectral lines seen in stars as being due to the Doppler effect. The effect is sometimes called the "Doppler-Fizeau effect". In 1868, British astronomer William Huggins was the first to determine the velocity.

In 1871, optical redshift was confirmed when the phenomenon was observed in Fraunhofer lines using solar rotation, about 0.1 Å in the red In 1901 Aristarkh Belopolsky verified optical redshift in the laboratory using a system of rotating mirrors.

The earliest occurrence of the term "red-shift" in print (in this hyphenated form), appears to be by American astronomer Walter S. Adams in 1908, where he mentions "Two methods of investigating that nature of the nebular red-shift" The word doesn't appear unhyphenated, perhaps indicating a more common usage of its German equivalent, Rotverschiebung, until about 1934 by Willem de Sitter.

Beginning with observations in 1912, Vesto Slipher discovered that most spiral nebulae had considerable redshifts. Subsequently, Edwin Hubble discovered an approximate relationship between the redshift of such "nebulae" (now known to be galaxies in their own right) and the distance to them with the formulation of his eponymous Hubble's law. These observations corroborated Alexander Friedman's 1922 work, in which he derived the famous Friedmann equations. They are today considered strong evidence for an expanding universe and the Big Bang theory.

Measurement, Characterization, and Interpretation

The spectrum of light that comes from a single source can be measured. To determine the redshift, features in the spectrum such as absorption lines, emission lines, or other variations in light intensity, are searched for. If found, these features can be compared with known features in the spectrum of various chemical compounds found in experiments where that compound is located on earth. A very common atomic element in space is hydrogen. The spectrum of originally featureless light shined through hydrogen will show a signature spectrum specific to hydrogen that has features at regular intervals. If restricted to absorption lines it would look similar to the illustration (top right). If the same pattern of intervals is seen in an observed spectrum from a distant source but occurring at shifted wavelengths, it can be identified as hydrogen too. If the same spectral line is identified in both spectra but at different wavelengths then the redshift can be calculated using the table below. Determining the redshift of an object in this way requires a frequency- or wavelength-range. In order to calculate the redshift one has to know the wavelength of the emitted light in the rest frame of the source, in other words, the wavelength that would be measured by an observer located adjacent to and comoving with the source. Since in astronomical applications this measurement cannot be done directly, because that would require travelling to the distant star of interest, the method using spectral lines described here is used instead. Redshifts cannot be calculated by looking at unidentified features whose rest-frame frequency is unknown, or with a spectrum that is featureless or white noise (random fluctuations in a spectrum).

Redshift (and blue shift) may be characterized by the relative difference between the observed and emitted wavelengths (or frequency) of an object. In astronomy, it is customary to refer to this change using a dimensionless quantity called z. If λ represents wavelength and f represents frequency (note, $\lambda f = c$ where c, is the speed of light), then z is defined by the equations:

Calculation of redshift, z

Based on wavelength

Based on frequency:

$$z = \frac{\lambda_{obsv} - \lambda_{emit}}{\lambda_{emit}} \qquad z = \frac{f_{emit} - f_{obsv}}{f_{obsv}}$$

$$1+z = \frac{\lambda_{obsv}}{\lambda_{emit}} \qquad 1+z = \frac{f_{emit}}{f_{obsv}}$$

After z is measured, the distinction between redshift and blue shift is simply a matter of whether z is positive or negative. See the mechanisms section below for some basic interpretations that follow when either a redshift or blue shift is observed. For example, Doppler effect blue shifts ($z < 0$) are associated with objects approaching (moving closer to) the observer with the light shifting to greater energies. Conversely, Doppler effect redshifts ($z > 0$) are associated with objects receding (moving away) from the observer with the light shifting to lower energies. Likewise, gravitational blue shifts are associated with light emitted from a source residing within a weaker gravitational field observed within a stronger gravitational field, while gravitational redshifting implies the opposite conditions.

Redshift Formulas

In general relativity one can derive several important special-case formulas for redshift in certain special spacetime geometries, as summarized in the following table. In all cases the magnitude of the shift (the value of z) is independent of the wavelength (*see Redshift summary*).

Doppler Effect

If a source of the light is moving away from an observer, then redshift ($z > 0$) occurs; if the source moves towards the observer, then blue shift ($z < 0$) occurs. This is true for all electromagnetic waves and is explained by the Doppler effect. Consequently, this type of redshift is called the Doppler redshift. If the source moves away from the

Redshift Summary

Redshift type	Geometry	Formula
Relativistic	Minkowski space	$1+z = \gamma\left(1+\frac{v_\parallel}{c}\right)$
Doppler	(flat spacetime)	$\left(z \approx \frac{v_\parallel}{c} \text{ for small} v\right)$
Cosmological redshift	FLRW spacetime (expanding Big Bang universe)	$1+z = \frac{a_{now}}{a_{then}}$
		$1+z = \sqrt{\frac{g_{tt}(\text{receiver})}{g_{tt}(\text{source})}}$
Gravitational	any stationary spacetime	(for the Schwarzschild geometry,
redshift	(e.g. the Schwarzschild geometry)	$1+z = \sqrt{\frac{1-\frac{2GM}{r_{receiver}}}{1-\frac{2GM}{r_{source}}}}$

observer with velocity υ, then, ignoring relativistic effects, the redshift is given by:

$$z \approx \frac{v}{c} \text{ (Since } \gamma \approx 1)$$

where c is the speed of light. In the classical Doppler effect, the frequency of the source is not modified, but the recessional motion causes the illusion of a lower frequency.

Relativistic Doppler Effect

A more complete treatment of the Doppler redshift requires considering relativistic effects associated with motion of sources close to the speed of light. A complete derivation of the effect can be found in the article on the relativistic Doppler effect (also see chapter 2 for relative doppler effect). In brief, objects moving close to the speed of light will experience deviations from the above formula due to the time dilation of special relativity which can be corrected for by introducing the Lorentz factor υ into the classical Doppler formula as follows:

$$1 + z = \left(1 + \frac{v}{c}\right)\gamma$$

This phenomenon was first observed in a 1938 experiment performed by Herbert E. Ives and G.R. Stilwell, called the Ives-Stilwell experiment.

Since the Lorentz factor is dependent only on the magnitude of the velocity, this causes the redshift associated with the relativistic correction to be independent of the orientation of the source movement. In contrast, the classical part of the formula is dependent on the projection of the movement of the source into the line-of-sight which yields different results for different orientations. Consequently, for an object moving at an angle θ to the observer (zero angle is directly away from the observer), the full form for the relativistic Doppler effect becomes:

$$1+z=\frac{1+v\cos(\theta)/c}{\sqrt{1-v^2/c^2}}$$

and for motion solely in the line of sight ($\theta = 0°$), this equation reduces to:

$$1+z=\sqrt{\frac{1+\frac{v}{c}}{1-\frac{v}{c}}}$$

For the special case that the source is moving at right angles ($\theta = 90°$) to the detector, the relativistic redshift is known as the transverse redshift, and a redshift:

$$1+z=\frac{1}{\sqrt{1-v^2/c^2}}$$

is measured, even though the object is not moving away from the observer. Even if the source is moving towards the observer, if there is a transverse component to the motion then there is some speed at which the dilation just cancels the expected blue shift and at higher speed the approaching source will be redshifted.

Expansion of Space

In the early part of the twentieth century, Slipher, Hubble and others made the first measurements of the redshifts and blue shifts of galaxies beyond the Milky Way. They initially interpreted these redshifts and blue shifts as due solely to the Doppler effect, but later Hubble discovered a rough correlation between the increasing redshifts and the increasing distance of galaxies. Theorists almost immediately realized that these observations could be explained by a different mechanism for producing redshifts. Hubble's law of the correlation between redshifts and distances is required by models of cosmology derived from general relativity that have a metric expansion of space. As a result, photons propagating through the expanding space are stretched, creating the cosmological redshift. This differs from the Doppler effect redshifts described above because the velocity boost (i.e. the Lorentz transformation) between the source and observer is not due to classical momentum and energy transfer, but instead the photons increase in wavelength and redshift as the space through which they are traveling expands. The observational consequences of this effect can be derived using the equations from general relativity that describe a homogeneous and isotropic universe.

To derive the redshift effect, use the geodesic equation for a light wave, which is

$$ds^2 = 0 = -c^2dt^2 + \frac{a^2\, dr^2}{1 - kr^2}$$

where

- ds is the Lorentzian line element
- dt is the time interval
- dr is the spatial interval
- c is the speed of light
- a is the time-dependent cosmic scale factor
- k is the curvature per unit area.

For an observer observing the crest of a light wave at a position r = 0 and time t = now, the crest of the light wave was emitted at a time t = then in the past and a distant position r = R. Integrating over the path in both space and time that the light wave travels yields:

$$c\int_{t_{\text{then}}}^{t_{\text{now}}} \frac{dt}{a} = \int_R^0 \frac{dr}{\sqrt{1-kr^2}}$$

In general, the wavelength of light is not the same for the two positions and times considered due to the changing properties of the metric. When the wave was emitted, it had a wavelength λthen. The next crest of the light wave was emitted at a time.

$$t = t_{then} + \lambda_{then} / c$$

The observer sees the next crest of the observed light wave with a wavelength λnow to arrive at a time

$$t = t_{now} + \lambda_{now} / c$$

Since the subsequent crest is again emitted from r = R and is observed at r = 0, the following equation can be written

$$c\int_{t_{then}+\lambda_{then}/c}^{t_{now}+\lambda_{now}/c} \frac{dt}{a} = \int_{R}^{0} \frac{dr}{\sqrt{1-kr^2}}$$

The right-hand side of the two integral equations above are identical which means

$$c\int_{t_{then}+\lambda_{then}/c}^{t_{now}+\lambda_{now}/c} \frac{dt}{a} = c\int_{t_{then}}^{t_{now}} \frac{dt}{a}$$

or, alternatively,

$$c\int_{t_{now}}^{t_{now}+\lambda_{now}/c} \frac{dt}{a} = \int_{t_{then}}^{t_{then}+\lambda_{then}/c} \frac{dt}{a}$$

For very small variations in time (over the period of one cycle of a light wave) the scale factor is essentially a constant ($a = a_{now}$ today and $a = a_{then}$ previously). This yields

$$\frac{t_{now} + \lambda_{now} / c}{a_{now}} - \frac{t_{now}}{a_{now}} = \frac{t_{then} + \lambda_{then} / c}{a_{then}} - \frac{t_{then}}{a_{then}}$$

which can be rewritten as

$$\frac{\lambda_{\text{now}}}{\lambda_{\text{then}}} = \frac{a_{\text{now}}}{a_{\text{then}}}$$

Using the definition of redshift provided above, the equation:

$$1 + z = \frac{a_{\text{now}}}{a_{\text{then}}}$$

is obtained. In an expanding universe such as the one we inhabit, the scale factor is monotonically increasing as time passes, thus, z is positive and distant galaxies appear redshifted.

This type of redshift is called the cosmological redshift or Hubble redshift. If the universe were contracting instead of expanding, we would see distant galaxies blue shifted by an amount proportional to their distance instead of redshifted

These galaxies are not receding simply by means of a physical velocity in the direction away from the observer; instead, the intervening space is stretching, which accounts for the large-scale isotropy of the effect demanded by the cosmological principle. For cosmological redshifts of $z < 0.01$ the effects of spacetime expansion are minimal and cosmological redshifts can be dominated by additional Doppler redshifts and blue shifts caused by the peculiar motions of the galaxies relative to one another. The difference between physical velocity and space expansion can be illustrated by the Expanding Rubber Sheet Universe, a common cosmological analogy used to describe the expansion of space. If two objects are represented by ball bearings and spacetime by a stretching rubber sheet, the Doppler effect is caused by rolling the balls across the sheet to create peculiar motion. The cosmological redshift occurs when the ball bearings are stuck to the sheet and the sheet is stretched. (Obviously, there are dimensional problems with the model, as the ball bearings should be in the sheet, and cosmological redshift produces higher velocities than Doppler does if the distance between two objects is large enough.)

In spite of the distinction between redshifts caused by the velocity of objects and the redshifts associated with the expanding universe, astronomers sometimes refer to "recession velocity" in the context of the redshifting of distant galaxies from the expansion of the Universe, even though it is only an apparent recession As a consequence, popular literature often uses the expression "Doppler redshift" instead of "cosmological redshift" to describe the motion of galaxies dominated by the expansion of spacetime, despite the fact that a "cosmological recessional speed" when calculated will not equal the velocity in the relativistic Doppler equation In particular, Doppler redshift is bound by special relativity; thus $\upsilon > c$ is impossible while, in contrast, $\upsilon > c$ is possible for cosmological redshift because the space which separates the objects (e.g., a quasar from the Earth) can expand faster than the speed of light. More mathematically, the viewpoint that "distant galaxies are receding" and the viewpoint that "the space between galaxies is expanding" are related by changing coordinate systems. Expressing this precisely requires working with the mathematics of the Friedmann-Robertson-Walker metric

Gravitational Redshift

In the theory of general relativity, there is time dilation within a gravitational well. This is known as the gravitational redshift or Einstein Shift. The theoretical derivation of this effect follows from the Schwarzschild solution of the Einstein equations which yields the following formula for redshift associated with a photon traveling in the gravitational field of an uncharged, nonrotating, spherically symmetric mass:

$$1 + z = \frac{1}{\sqrt{1 - \left(\frac{2GM}{rc^2}\right)}}$$

where

- G is the gravitational constant;
- M is the mass of the object creating the gravitational field;
- r is the radial coordinate of the observer (which is analogous to the classical distance from the center of the object, but is actually a Schwarzschild coordinate); and
- c is the speed of light.

This gravitational redshift result can be derived from the assumptions of special relativity and the equivalence principle; the full theory of general relativity is not required.

The effect is very small but measurable on Earth using the Mössbauer effect and was first observed in the Pound-Rebka experiment. However, it is significant near a black hole, and as an object approaches the event horizon the red shift becomes infinite. It is also the dominant cause of large angular-scale temperature fluctuations in the cosmic microwave background radiation.

Observations in Astronomy

The redshift observed in astronomy can be measured because the emission and absorption spectra for atoms are distinctive and well known, calibrated from spectroscopic experiments in laboratories on Earth. When the redshift of various absorption and emission lines from a single astronomical object is measured, z is found to be remarkably constant. Although distant objects may be slightly blurred and lines broadened, it is by no more than can be explained by thermal or mechanical motion of the source. For these reasons and others, the consensus among astronomers is that the redshifts they observe are due to some combination of the three established forms of Doppler-like redshifts. Alternative hypotheses are not generally considered plausible.

Spectroscopy, as a measurement, is considerably more difficult than simple photometry, which measures the brightness of astronomical objects through certain filters When photometric data are all that is available (for example, the Hubble Deep Field and the Hubble Ultra Deep Field), astronomers rely on a technique for measuring photometric redshifts. Due to the filter being sensitive to a range of wavelengths and the technique relying on making many assumptions about the nature of the spectrum at the light-source, errors for these sorts of measurements can range up to $\varepsilon z = 0.5$, and are much less reliable than spectroscopic determinations. However, photometry does allow at least for a qualitative characterization of a redshift. For example, if a sun-like spectrum had a redshift of $z = 1$, it would be brightest in the infrared rather than at the yellow-green color associated with the peak of its blackbody spectrum, and the light intensity will be reduced in the filter by a factor of two $(1+z)$.

Local Observations

In nearby objects (within our Milky Way galaxy) observed redshifts are almost always related to the line-of-sight velocities associated with the objects being observed. Observations of such redshifts and blue shifts have enabled astronomers to measure velocities and parametrize the masses of the orbiting stars in spectroscopic binaries, a method first employed in 1868 by British astronomer William Huggins Similarly, small redshifts and blue shifts detected in the spectroscopic measurements of individual stars are one way astronomers have been able to diagnose and measure the presence and characteristics of planetary systems around other stars. Measurements of redshifts to fine detail are used in helioseismology to determine the precise movements of the photosphere of the Sun. Redshifts have also been used to make the first measurements of the rotation rates of planets, velocities of interstellar clouds, the rotation of galaxies, and the dynamics of accretion onto neutron stars and black holes which exhibit both Doppler and gravitational redshifts. Additionally, the temperatures of various emitting and absorbing objects can be obtained by measuring Doppler broadening—effectively redshifts and blue shifts over a single emission or absorption line. By measuring the broadening and shifts of the 21-centimetre hydrogen line in different directions, astronomers have been able to measure the recessional velocities of interstellar gas, which in turn reveals the rotation curve of our Milky Way Similar measurements have been performed on other galaxies, such as Andromeda. As a diagnostic tool, redshift measurements are one of the most important spectroscopic measurements made in astronomy.

Extragalactic Observations

The most distant objects exhibit larger redshifts corresponding to the Hubble flow of the universe. The largest observed redshift, corresponding to the greatest distance and furthest back in time, is that of the cosmic microwave background radiation; the numerical value of its redshift is about z = 1089 (z = 0 corresponds to present time), and it shows the state of the Universe about 13.7 billion years ago, and 379,000 years after the initial moments of the Big Bang.

The luminous point-like cores of quasars were the first "high-red shift" (z > 0.1) objects discovered before the improvement of telescopes allowed for the discovery of other high-redshift galaxies.

For galaxies more distant than the Local Group and the nearby Virgo Cluster, but within a thousand megaparsecs or so, the redshift is approximately proportional to the galaxy's distance. This correlation was first observed by Edwin Hubble and has come to be known as Hubble's law. Vesto Slipher was the first to discover galactic redshifts, in about the year 1912, while Hubble correlated Slipher's measurement with distances he measured by other means to formulate his Law. In the widely accepted cosmological model based on general relativity, redshift is mainly a result of the expansion of space: this means that the farther away a galaxy is from us, the more the space has expanded in the time since the light left that galaxy, so the more the light has been stretched, the more redshifted the light is, and so the faster it appears to be moving away from us. Hubble's law follows in part from the Copernican principle. Because it is usually not known how luminous objects are, measuring the redshift is easier than more direct distance measurements, so redshift is sometimes in practice converted to a crude distance measurement using Hubble's law.

Gravitational interactions of galaxies with each other and clusters cause a significant scatter in the normal plot of the Hubble diagram. The peculiar velocities associated with galaxies superimpose a rough trace of the mass of virialized objects in the universe. This effect leads to such phenomena as nearby galaxies (such as the Andromeda Galaxy) exhibiting blue shifts as we fall towards a common barycenter, and redshift maps of clusters showing a Finger of God effect due to the scatter of peculiar velocities in a roughly spherical distribution. This added component gives cosmologists a chance to measure the masses of objects independent of the mass to light ratio (the ratio of a galaxy's mass in solar masses to its brightness in solar luminosities), an important tool for measuring dark matter.

The Hubble law's linear relationship between distance and redshift assumes that the rate of expansion of the universe is constant. However, when the universe was much younger, the expansion rate, and thus the Hubble "constant", was larger than it is today. For more distant galaxies, then, whose light has been travelling to us for much longer times, the approximation of constant expansion rate fails, and the Hubble law becomes a non-linear integral relationship and dependent on the history of the expansion rate since the emission of the light from the galaxy in question. Observations of the redshift-distance relationship can be used, then, to determine the expansion history of the universe and thus the matter and energy content.

While it was long believed that the expansion rate has been continuously decreasing since the Big Bang, recent observations of the redshift-distance relationship using Type Ia supernovae have suggested that in comparatively recent times the expansion rate of the universe has begun to accelerate.

Highest Redshifts

Currently, the objects with the highest known redshifts are galaxies. The most reliable redshifts are from spectroscopic data, and the highest confirmed spectroscopic redshift of a galaxy is that of IOK-1 at a redshift z = 6.96. Slightly less reliable are Lyman-break redshifts, the highest of which is the lensed galaxy A1689-zD1 at a redshift z = 7.6 and the next highest being z = 7.0 while as-yet unconfirmed from a gravitational lens observed in a distant galaxy cluster may indicate a galaxy with a redshift of z = 10.1.

The most distant observed gamma ray burst is GRB 080913, which had a redshift of 6.7.

The most distant known quasar, CFHQS J2329-0301, is at z = 6.43 The highest known redshift radio galaxy (TN J0924-2201) is at a redshift z = 5.2 and the highest known redshift molecular material is the detection of emission from the CO molecule from the quasar SDSS J1148+5251 at z = 6.42

Redshift Surveys

With the advent of automated telescopes and improvements in spectroscopes, a number of collaborations have been made to map the universe in redshift space. By combining redshift with angular position data, a redshift survey maps the 3D distribution of matter within a field of the sky. These observations are used to measure properties of the large-scale structure of the universe. The Great Wall, a vast supercluster of galaxies over 500 million light-years wide, provides a dramatic example of a large-scale structure that redshift surveys can detect.

The first redshift survey was the CfA Redshift Survey, started in 1977 with the initial data collection completed in 1982. More recently, the 2dF Galaxy Redshift Survey determined the large-scale structure of one section of the Universe, measuring z-values for over 220,000 galaxies; data collection was completed in 2002, and the final data set was released 30 June 2003. (In addition to mapping large-scale patterns of galaxies, 2dF established an upper limit on neutrino mass.) Another notable investigation, the Sloan Digital Sky Survey (SDSS), is ongoing as of 2005 and aims to obtain measurements on around 100 million objects. SDSS has recorded redshifts for galaxies as high as 0.4, and has been involved in the detection of quasars beyond z = 6. The DEEP2 Redshift Survey uses the Keck telescopes with the new "DEIMOS" spectrograph; a follow-up to the pilot program DEEP1, DEEP2 is designed to measure faint galaxies with redshifts 0.7 and above, and it is therefore planned to provide a complement to SDSS and 2dF.

Effects due to Physical Optics or Radiative Transfer

The interactions and phenomena summarized in the subjects of radiative transfer and physical optics can result in shifts in the wavelength and frequency of electromagnetic radiation. In such cases the shifts correspond to a physical energy transfer to matter or other photons rather than being due to a transformation between reference frames. These shifts can be due to such physical phenomena as coherence effects or the scattering of electromagnetic radiation whether from charged elementary particles, from particulates, or from fluctuations of the index of refraction in a dielectric medium as occurs in the radio phenomenon of radio whistlers While such phenomena are sometimes referred to as "redshifts" and "blue shifts", the physical interactions of the electromagnetic radiation field with itself or intervening matter distinguishes these phenomena from the reference-frame effects. In astrophysics, light-matter interactions that result in energy shifts in the radiation field are generally referred to as "reddening" rather than "redshifting" which, as a term, is normally reserved for the effects discussed above.

In many circumstances scattering causes radiation to redden because entropy results in the predominance of many low-energy photons over few high-energy ones (while conserving total energy) Except possibly under carefully controlled conditions, scattering does not produce the same relative change in wavelength across the whole spectrum; that is, any calculated z is generally a function of wavelength. Furthermore, scattering from random media generally occurs at many angles, and z is a function of the scattering angle. If multiple scattering occurs, or the scattering particles have relative motion, then there is generally distortion of spectral lines as well.

In interstellar astronomy, visible spectra can appear redder due to scattering processes in a phenomenon referred to as interstellar reddening — similarly Rayleigh scattering causes the atmospheric reddening of the Sun seen in the sunrise or sunset and causes the rest of the sky to have a blue color. This phenomenon is distinct from redshifting because the spectroscopic lines are not shifted to other wavelengths in reddened objects and there is an additional dimming and distortion associated with the phenomenon due to photons being scattered in and out of the line-f-sight.

SPECIAL RELATIVITY

Special relativity (SR) (also known as the special theory of relativity or STR) is the physical theory of measurement in inertial frames of reference proposed in 1905 by Albert Einstein (after the considerable and independent contributions of Hendrik Lorentz and Henri Poincaré and others) in the paper "On the Electrodynamics of Moving Bodies". It generalizes Galileo's principle of relativity-that all uniform motion is relative, and that there is no absolute and well-defined state of rest (no privileged reference frames)-from mechanics to all the laws of physics, including both the laws of mechanics and of electrodynamics, whatever they may be. Special relativity incorporates the principle that the speed of light is the same for all inertial observers regardless of the state of motion of the source.

This theory has a wide range of consequences which have been experimentally verified including counter-intuitive ones such as length contraction, time dilation and relativity of simultaneity, contradicting the classical notion that the duration of the time interval between two events is equal for all observers. (On the other hand, it introduces the space-time interval, which is invariant.) Combined with other laws of physics, the two postulates of special relativity predict the equivalence of matter and energy, as expressed in the mass-energy equivalence formula $E = mc^2$, where c is the speed of light in a vacuum. The predictions of special relativity agree well with Newtonian mechanics in their common realm of applicability, specifically in experiments in which all velocities are small compared to the speed of light.

The theory is termed "special" because it applies the principle of relativity only to inertial frames. Einstein developed general relativity to apply the principle more generally, that is, to any frame so as to handle general coordinate transformations, and that theory includes the effects of gravity. From the theory of general relativity it follows that, in the absence of a gravitational field, special relativity will still apply locally to observers moving on arbitrary trajectories.

Special relativity reveals that c is not just the velocity of a certain phenomenon, namely the propagation of electromagnetic radiation (light)—but rather a fundamental feature of the way space and time are unified as spacetime. A consequence of this is that it is impossible for any particle that has mass to be accelerated to the speed of light.

Postulates

In his autobiographical notes published in November 1949 Einstein described how he had arrived at the two fundamental postulates on which he based the special theory of relativity. After describing in detail the state of both mechanics and electrodynamics at the beginning of the 20th century, he wrote:

> "Reflections of this type made it clear to me as long ago as shortly after 1900, i.e., shortly after Planck's trailblazing work, that neither mechanics nor electrodynamics could (except in limiting cases) claim exact validity. Gradually I despaired of the possibility of discovering the true laws by means of constructive efforts based on known facts. The longer and the more desperately I tried, the more I came to the conviction that only the discovery of a universal formal principle could lead us to assured results... How, then, could such a universal principle be found?

He discerned two fundamental propositions that seemed to be the most assured, regardless of the exact validity of either the (then) known laws of mechanics or electrodynamics. These propositions were:

1. the constancy of the velocity of light; and
2. the independence of physical laws (especially the constancy of the velocity of light) from the choice of inertial system.

The Principle of Relativity - The laws by which the states of physical systems undergo change are not affected, whether these changes of state be referred to the one or the other of two systems in uniform translatory motion relative to each other.

The Principle of Invariant Light Speed - Light in vacuum propagates with the speed c (a fixed constant) in terms of any system of inertial coordinates, regardless of the state of motion of the light source.

It should be noted that the derivation of special relativity depends not only on these two explicit postulates, but also on several tacit assumptions (which are made in almost all theories of physics), including the isotropy and homogeneity of space and the independence of measuring rods and clocks from their past history.

Following Einstein's original presentation of special relativity in 1905, many different sets of postulates have been proposed in various alternative derivations. However, the most common set of postulates remains those employed by Einstein in his original paper. A more mathematical statement of the Principle of Relativity made later by Einstein, which introduces the concept of simplicity not mentioned above is the two postulates of special relativity imply the applicability to physical laws of the Poincaré group of symmetry transformations, of which the Lorentz transformations are a subset, thereby providing a mathematical framework for special relativity. Many of Einstein's papers present derivations of the Lorentz transformation based upon these two principles.

Einstein consistently based the derivation of Lorentz invariance (the essential core of special relativity) on just the two basic principles of relativity and light-speed invariance. He wrote:

> "The insight fundamental for the special theory of relativity is this: The assumptions relativity and light speed invariance are compatible if relations of a new type ("Lorentz transformation") are postulated for the conversion of coordinates and times of events ... The universal principle of the special theory of relativity is contained in the postulate: The laws of physics are invariant with respect to Lorentz transformations (for the transition from one inertial system to any other arbitrarily chosen inertial system). This is a restricting principle for natural laws...

Thus many modern treatments of special relativity base it on the single postulate of universal Lorentz covariance, or, equivalently, on the single postulate of Minkowski spacetime.

Mass-energy Equivalence

In addition to the papers referenced above—which give derivations of the Lorentz transformation and describe the foundations of special relativity—Einstein also wrote at least four papers giving heuristic arguments for the equivalence (and transmutability) of mas and energy (the famous formula $E = mc^2$).

Mass-energy equivalence does not follow from the two basic postulates of special relativity by themselves. The first of Einstein's papers on this subject was Does the Inertia of a Body Depend upon its Energy Content? in 1905 In this first paper and in each of his subsequent three papers on this subject Einstein augmented the two fundamental principles by postulating the relations involving momentum and energy of electromagnetic waves implied by Maxwell's equations (the assumption of which, of course, entails among other things the assumption of the constancy of the speed of light). It has been suggested that Einstein's original argument was fallacious. Other authors suggest that the argument was merely inconclusive by virtue of some implicit assumptions lacking experimental verification at the time.

Einstein acknowledged in his 1907 survey paper on special relativity that it was problematic to rely on Maxwell's equations for the heuristic mass-energy argument.

Lack of an Absolute Reference Frame

The principle of relativity, which states that there is no preferred inertial reference frame, dates back to Galileo, and was incorporated into Newtonian Physics. However, in the late 19th century, the existence of electromagnetic waves led physicists to suggest that the universe was filled with a substance known as "aether", which would act as the medium through which these waves, or vibrations traveled. The aether was thought to constitute an absolute reference frame against which speeds could be measured. In other words, the aether was the only fixed or motionless thing in the universe. Aether supposedly had some wonderful properties: it was sufficiently elastic that it could support electromagnetic waves, and those waves could interact with matter, yet it offered no resistance to bodies passing through it. The results of various experiments, including the Michelson-Morley experiment, indicated that the Earth was always 'stationary' relative to the aether–something that was difficult to explain, since the Earth is in orbit around the Sun. Einstein's elegant solution was to discard the notion of an aether and an absolute state of rest. Special relativity is formulated so as to not assume that any particular frame of reference is special; rather, in relativity, any reference frame moving with uniform motion will observe the same laws of physics. In particular, the speed of light in a vacuum is always measured to be c, even when measured by multiple systems that are moving at different (but constant) velocities.

Consequences

Einstein has said that all of the consequences of special relativity can be derived from examination of the Lorentz transformations.

These transformations, and hence special relativity, lead to different physical predictions than Newtonian mechanics when relative velocities become comparable to the speed of light. The speed of light is so much larger than anything humans encounter that some of the effects predicted by relativity are initially counter-intuitive:

- *Time dilation*—the time lapse between two events is not invariant from one observer to another, but is dependent on the relative speeds of the observers' reference frames (e.g., the twin paradox which concerns a twin who flies off in a spaceship traveling near the speed of light and returns to discover that his or her twin sibling has aged much more).
- *Relativity of simultaneity*—two events happening in two different locations that occur simultaneously in the reference frame of one inertial observer, may occur non-simultaneously in the reference frame of another inertial observer (lack of absolute simultaneity).
- *Lorentz contraction*—the dimensions (e.g., length) of an object as measured by one observer may be smaller than the results of measurements of the same object made by another observer (e.g., the ladder paradox involves a long ladder traveling near the speed of light and being contained within a smaller garage).
- *Composition of velocities*—velocities (and speeds) do not simply 'add', for example if a rocket is moving at ? the speed of light relative to an observer, and the rocket fires a missile at the speed of light relative to the rocket, the missile does not exceed the speed of light relative to the observer. (In this example, the observer would see the missile travel with a speed of 12/13 the speed of light).

- *Inertia and momentum*—as an object's speed approaches the speed of light from an observer's point of view, its mass appears to increase thereby making it more and more difficult to accelerate it from within the observer's frame of reference.
- *Equivalence of mass* and *energy*, $E = mc^2$—The energy content of an object at rest with mass m equals mc^2. Conservation of energy implies that in any reaction a decrease of the sum of the masses of particles must be accompanied by an increase in kinetic energies of the particles after the reaction. Similarly, the mass of an object can be increased by taking in kinetic energies.

Reference Frames, Coordinates and the Lorentz Transformation

Relativity theory depends on "reference frames". A reference frame is an observational perspective in space at rest, or in uniform motion, from which a position can be measured along 3 spatial axes. In addition, a reference frame has the ability to determine measurements of the time of events using a 'clock' (any reference device with uniform periodicity).

An event is an occurrence that can be assigned a single unique time and location in space relative to a reference frame: it is a "point" in space-time. Since the speed of light is constant in relativity in each and every reference frame, pulses of light can be used to unambiguously measure distances and refer back the times that events occurred to the clock, even though light takes time to reach the clock after the event has transpired.

For example, the explosion of a firecracker may be considered to be an "event". We can completely specify an event by its four space-time coordinates: The time of occurrence and its 3-dimensional spatial location define a reference point. Let's call this reference frame S.

In relativity theory we often want to calculate the position of a point from a different reference point.

Suppose we have a second reference frame S′, whose spatial axes and clock exactly coincide with that of S at time zero, but it is moving at a constant velocity with respect to S along the-axis.

Since there is no absolute reference frame in relativity theory, a concept of 'moving' doesn't strictly exist, as everything is always moving with respect to some other reference frame. Instead, any two frames that move at the same speed in the same direction are said to be comoving. Therefore S and S′ are not comoving.

Let's define the event to have space-time coordinates (t', x', y', z') in system S and (t', x', y', z') in S′. Then the Lorentz transformation specifies that these coordinates are related in the following way:

$$t' = \gamma \left(t - \frac{vx}{c^2}\right)$$

$$x' = \gamma\ (x - vt)$$

$$y' = y$$

$$z' = z$$

where

$$\gamma = \frac{1}{\sqrt{1 - v^2/c^2}}$$

is called the Lorentz factor and C is the speed of light in a vacuum.

The Y and Z coordinates are unaffected, only the X and T axes transformed. These Lorentz transformations form a one-parameter group of linear mappings, that parameter being called rapidity.

A quantity invariant under Lorentz transformations is known as a Lorentz scalar.

Simultaneity

From the first equation of the Lorentz transformation in terms of coordinate differences

$$\Delta t' = \gamma \left(\Delta t' - \frac{v\Delta x}{c^2}\right)$$

it is clear that two events that are simultaneous in frame S (satisfying $\Delta t = 0$ are not necessarily simultaneous in another inertial frame S′ (satisfying, $\Delta t' = 0$). Only if these events are colocal in frame S (satisfying $\Delta x = 0$), will they be simultaneous in another frame S′.

Time Dilation and Length Contraction

Writing the Lorentz transformation and its inverse in terms of coordinate differences we get

$$\begin{cases} \Delta t' = \gamma \left(\Delta t' - \dfrac{v\Delta_x}{c^2}\right) \\ \Delta t' = \gamma \left(\Delta x' - v\Delta t\right) \end{cases}$$

and

$$\begin{cases} \Delta t' = \gamma \left(\Delta t' - \dfrac{v\Delta_x{}'}{c^2}\right) \\ \Delta x' = \gamma \left(\Delta x' - v\Delta t'\right) \end{cases}$$

Suppose we have a clock at rest in the unprimed system S. Two consecutive ticks of this clock are then characterized by $\Delta x = 0$. If we want to know the relation between the times between these ticks as measured in both systems, we can use the first equation and find:

$\Delta t' = \gamma \Delta t$ (for events satisfying $\Delta x = 0$)

This shows that the time Δt′ between the two ticks as seen in the 'moving' frame S′ is larger than the time Δt between these ticks as measured in the rest frame of the clock. This phenomenon is called time dilation.

Similarly, suppose we have a measuring rod at rest in the unprimed system. In this system, the length of this rod is written as Δx. If we want to find the length of this rod as measured in the 'moving' system S′, we must make sure to measure the distances x′ to the end points of the rod simultaneously in the primed frame S′. In other words, the measurement is characterized by Δt = 0 which we can combine with the fourth equation to find the relation between the lengths Δx $\Delta x'$:

$\Delta x' = \frac{\Delta x}{\gamma}$ (for events satisfying $\Delta t' = 0$)

This shows that the length $\Delta x'$ of the rod as measured in the 'moving' frame S′ is shorter than the length Δx in its own rest frame. This phenomenon is called length contraction or Lorentz contraction.

These effects are not merely appearances; they are explicitly related to our way of measuring time intervals between events which occur at the same place in a given coordinate system (called "co-local" events). These time intervals will be different in another coordinate system moving with respect to the first, unless the events are also simultaneous. Similarly, these effects also relate to our measured distances between separated but simultaneous events in a given coordinate system of choice. If these events are not co-local, but are separated by distance (space), they will not occur at the same spacial distance from each other when seen from another moving coordinate system.

Causality and Prohibition of Motion Faster than Light

There is a frame of reference in which event A and event B occur at the same location in space, separated only by occurring at different times. If A precedes B in that frame, then A precedes B in all frames. It is hypothetically possible for matter (or information) to travel from A to B, so there can be a causal relationship (with A the cause and B the effect).

The interval AC in the diagram is 'space-like'; i.e., there is a frame of reference in which event A and event C occur simultaneously, separated only in space. However there are also frames in which A precedes C and frames in which C precedes A. If it were possible for a cause-and-effect relationship to exist between events A and C, then paradoxes of causality would result. For example, if A was the cause, and C the effect, then there would be frames of reference in which the effect preceded the cause. Although this in itself won't give rise to a paradox, one can show that faster than light signals can be sent back into one's own past. A causal paradox can then be constructed by sending the signal if and only if no signal was received previously.

Therefore, one of the consequences of special relativity is that (assuming causality is to be preserved), no information or material object can travel faster than light. On the other hand, the logical situation is not as clear in the case of general relativity, so it is an open question whether there is some fundamental principle that preserves causality (and therefore prevents motion faster than light) in general relativity.

Even without considerations of causality, there are other strong reasons why faster-than-light travel is forbidden by special relativity. For example, if a constant force is applied to an object for a limitless amount of time, then integrating $F = dp/dt$ gives a momentum that grows without bound, but this is simply because $p = m\upsilon v$ approaches infinity as v approaches c. To an observer who is not accelerating, it appears as though the object's inertia is increasing, so as to produce a smaller acceleration in response to the same force. This behavior is in fact observed in particle accelerators.

Composition of Velocities

If the observer in S sees an object moving along the x axis at velocity w, then the observer in the S′ system, a frame of reference moving at velocity v in the x direction with respect to S, will see the object moving with velocity w′ where

$$w' = \frac{w - v}{1 - wv/c^2}$$

This equation can be derived from the space and time transformations above

$$w' = \frac{dx'}{dt'} = \frac{\gamma\,(dx - v dt)}{\gamma\,(dt - v dx/c^2)} = \frac{(dx/dt) - v}{1 - (v/c^2)\,(dx/dt)}$$

Notice that if the object were moving at the speed of light in the S system (i.e. w = c), then it would also be moving at the speed of light in the S′ system. Also, if both w and v are small with respect to the speed of light, we will recover the intuitive Galilean transformation of velocities:

$$w' \approx w - v$$

Relativistic Mechanics

In addition to modifying notions of space and time, special relativity forces one to reconsider the concepts of mass, momentum, and energy, all of which are important constructs in Newtonian mechanics. Special relativity shows, in fact, that these concepts are all different aspects of the same physical quantity in much the same way that it shows space and time to be interrelated.

There are a couple of (equivalent) ways to define momentum and energy in SR. One method uses conservation laws. If these laws are to remain valid in SR they must be true in every possible reference frame. However, if one does some simple thought experiments using the Newtonian definitions of momentum and energy, one sees that these quantities are not conserved in SR. One can rescue the idea of conservation by making some small modifications to the definitions to account for relativistic velocities. It is these new definitions which are taken as the correct ones for momentum and energy in SR.

The energy and momentum of an object with invariant mass m (also called rest mass in the case of a single particle), moving with velocity v with respect to a given frame of reference, are given by

$$E = \gamma mc^2$$

$$p = \gamma mv$$

respectively, where γ (the Lorentz factor) is given by

$$\gamma = \frac{1}{\sqrt{1-(v/c)^2}}$$

The quantity γm is often called the relativistic mass of the object in the given frame of reference, although recently this concept is falling in disuse, and Lev B. Okun suggested that "this terminology . . . has no rational justification today", and should no longer be taught. Other physicists, including Wolfgang Rindler and T.R. Sandin, have argued that relativistic mass is a useful concept and there is little reason to stop using it. See Mass in special relativity for more information on this debate. Some authors use the symbol m to refer to relativistic mass, and the symbol m0 to refer to rest mass.

The energy and momentum of an object with invariant mass m are related by the formulas

$$E^2 - (pc)^2 = (mc^2)^2$$

$$pc^2 = Ev.$$

The first is referred to as the relativistic energy-momentum equation. While the energy E and the momentum **p** depend on the frame of reference in which they are measured, the quantity $E^2 - (pc)^2$ is invariant, being equal to the squared invariant mass of the object (up to the multiplicative constant c^4).

It should be noted that the invariant mass of a system

$$m_{tot} = \frac{\sqrt{E^2_{tot} - (p_{tot}c)^2}}{c^2}$$

is greater than the sum of the rest masses of the particles it is composed of (unless they are all stationary with respect to the center of mass of the system, and hence to each other). The sum of rest masses is not even always conserved in closed systems, since rest mass may be converted to particles which individually have no mass, such as photons. Invariant mass, however, is conserved and invariant for all observers, so long as the system remains closed. This is due to the fact that even massless particles contribute invariant mass to systems, as also does the kinetic energy of particles. Thus, even under transformations of rest mass to photons or kinetic energy, the invariant mass of a system which contains these energies still reflects the invariant mass associated with them.

Mass-energy Equivalence

For massless particles, m is zero. The relativistic energy-momentum equation still holds, however, and by substituting m with 0, the relation E = pc is obtained; when substituted into $Ev = c^2p$, it gives v = c: massless particles (such as photons) always travel at the speed of light.

A particle which has no rest mass (for example, a photon) can nevertheless contribute to the total invariant mass of a system, since some or all of its momentum is canceled by another particle, causing a contribution to the system's invariant mass due to the photon's energy. For single photons this does not happen, since the energy and momentum terms exactly cancel.

Looking at the above formula for invariant mass of a system, one sees that, when a single massive object is at rest (**v** = 0, **p** = 0), there is a non-zero mass remaining: $m_{rest} = E/c^2$. The corresponding energy, which is also the total energy when a single particle is at rest, is referred to as "rest energy". In systems of particles which are seen from a moving inertial frame, total energy increases and so does momentum. However, for single particles the rest mass remains constant, and for systems of particles the invariant mass remain constant, because in both cases, the energy and momentum increases subtract from each other, and cancel. Thus, the invariant mass of systems of particles is a calculated constant for all observers, as is the rest mass of single particles.

The Mass of Systems and Conservation of Invariant Mass

The inertial frame in which the momenta of all particles sums to zero is called the center of momentum frame. In this special frame, the relativistic energy-momentum equation has ρ = 0, and thus gives the mass of the system as merely the total energy of all parts of the system, divided by c^2

$$m = \sum E/c^2$$

This is the mass of any system which is measured in a frame where it has zero total momentum, such as a bottle of hot gas on a scale. In such a system, the mass which the scale weighs is the invariant mass, and it depends on the total energy of the system. It is thus more than the sum of the rest masses of the molecules, but also includes all the totaled energies in the system as well. Like energy and momentum, the invariant mass of closed systems cannot be changed so long as the system is closed, because the total relativistic energy of the system remains constant so long as nothing can enter or leave it.

An increase in the energy of such a system which is caused by translating the system to an inertial frame which is not the center of momentum frame, causes an increase in energy and momentum without an increase in invariant mass. Einstein's famous $E = mc^2$ equation, however, applies only to closed systems in their centre-of-momentum frame where momentum sums to zero.

Taking this formula at face value, we see that in relativity, mass is simply another form of energy. In 1927 Einstein remarked about special relativity: Under this theory mass is not an unalterable magnitude, but a magnitude dependent on (and, indeed, identical with) the amount of energy.

Einstein was not referring to closed systems in this remark, however. For, even in his 1905 paper, which first derived the relationship between mass and energy, Einstein showed that the energy of an object had to be increased for its invariant mass (rest mass) to increase. In such cases, the system is not closed (in Einstein's thought experiment, for example, a mass gives off two photons, which are lost).

Closed (Isolated Systems)

In a closed system the total energy, the total momentum, and hence the total invariant mass are conserved. Einstein's formula for invariant mass translates to its simplest $\Delta E = \Delta mc^2$ form, however, in non-closed systems in which energy is allowed to escape (for example, as heat and light). Einstein's equation shows that such systems must lose mass, in accordance with the above formula, in proportion to the energy they lose to the surroundings. Conversely, if one can measure the differences in mass between a system before it undergoes a reaction which releases heat and light, and the system after the reaction when heat and light have escaped, one can estimate the amount of energy which escapes the system. In both nuclear and chemical reactions, such energy represents the difference in binding energies of electrons in atoms (for chemistry) or between nucleons in nuclei (in atomic reactions). In both cases, the mass difference between reactants and (cooled) products measures the mass of heat and light which will escape the reaction, and thus (using the equation) give the equivalent energy of heat and light which may be emitted if the reaction proceeds.

In chemistry, the mass differences associated with the emitted energy are too small to measure. However, in nuclear reactions the energies are so large that they are associated with mass differences, which can be estimated in advance, if the products and reactants have been weighed (atoms can be weighed indirectly by using atomic masses, which are always the same for each nuclide). Thus, Einstein's formula becomes important when one has measured the masses of different atomic nuclei. By looking at the difference in masses, one can predict which nuclei have stored energy that can be released by certain nuclear reactions, providing important information which was useful in the development of nuclear energy and, consequently, the nuclear bomb. Historically, for example, Lise Meitner was able to use the mass differences in nuclei to estimate that there was enough energy available to make nuclear fission a favorable process. The implications of this formula on 20th-century life have made it one of the most famous equations in all of science.

Because the $E = mc^2$ equation applies to systems only in their center of momentum frame, it has been popularly misunderstood to mean that mass may be converted to energy, after which the mass disappears. This is incorrect, as mass never disappears in the center of momentum frame, since this type of mass is invariant mass, and is thus conserved unless it is allowed to escape. Instead, this equation, in context, means only that when any energy is added to, or escapes from, a system in the center-of-momentum frame, the system will be measured as having gained or lost mass, in proportion to the energy added or removed. Thus, in theory, if even an atomic bomb were placed in a box strong enough to hold its blast, and detonated upon a scale, the mass of this closed system would not change, and the scale would not move. Only when a transparent "window" was opened in the super-strong plasma-filled box, and light and heat were allowed to escape in a beam, and the bomb components to cool, would the system lose the mass associated with the energy of the blast. In a 21 kiloton bomb, for example, about a gram of light and heat is created. If this heat and light were allowed to escape, the remains of the bomb would lose a gram of mass, as it cooled. However, invariant mass cannot be destroyed in special relativity, but only moved from place to place. In this thought-experiment, the light and heat carry away the gram of mass, and would therefore deposit this gram of mass in the objects that absorb them.

Force

In special relativity, Newton's second law does not hold in its form **F** = m**a**, but it does if it is expressed as

$$F = \frac{dp}{dt}$$

where **p** is the momentum as defined above ($\mathbf{p} = \gamma m v$) and "m" is the invariant mass. Thus, the force is given by

$$F = m\frac{d(\gamma v)}{dt} = m\left(\frac{d\gamma}{(dt)}v + \gamma\ \frac{d\gamma}{(dt)}\right)$$

Carrying out the derivatives gives

$$F = \frac{\gamma^3 m v}{c^2}\ \frac{dv}{dt}v + \gamma m\mathbf{a}$$

which, taking into account the identity

$v\frac{dv}{dt} = \mathbf{v.a}$ can also be expressed as

$$\mathbf{F} = \frac{\gamma^3 m\,(\mathbf{v.a})}{c^2} v + \gamma m\mathbf{a}.$$

If the acceleration is separated into the part parallel to the velocity and the part perpendicular to it, one gets

$$\mathbf{F} = \frac{\gamma^3 m v^2}{c^2}\mathbf{a}_{||} + \gamma m\,(a_{||} + a_{\perp})$$

$$= \gamma^3 m\left(\frac{v^2}{c^2} + \frac{1}{\gamma^2}\right)a_{||} + \gamma m a_{\perp}$$

$$= \gamma^3 m\left(\frac{v^2}{c^2} + 1 - \frac{v^2}{c^2}\right)a_{||} + \gamma m a_{\perp}$$

$$= \gamma^3 m a_{||} + \gamma m a_{\perp}$$

Consequently in some old texts, γ^3m is referred to as the longitudinal mass, and γm is referred to as the transverse mass, which is the same as the relativistic mass.

Status

Special relativity is accurate only when gravitational potential is much less than c^2; in a strong gravitational field one must use general relativity (which becomes special relativity at the limit of weak field). At very small scales, such as at the Planck length and below, quantum effects must be taken into consideration resulting in quantum gravity. However, at macroscopic scales and in the absence of strong gravitational fields, special relativity is experimentally tested to extremely high degree of accuracy (10-20) and thus accepted by the physics community. Experimental results which appear to contradict it are not reproducible and are thus widely believed to be due to experimental errors.

Special relativity is mathematically self-consistent, and it is an organic part of all modern physical theories, most notably quantum field theory, string theory, and general relativity (in the limiting case of negligible gravitational fields).

Newtonian mechanics mathematically follows from special relativity at small velocities (compared to the speed of light) - thus Newtonian mechanics can be considered as a special relativity of slow moving bodies. See Status of special relativity for a more detailed discussion.

A few key experiments can be mentioned that led to special relativity:

- The *Trouton–Noble experiment* showed that the torque on a capacitor is independent on position and inertial reference frame–such experiments led to the first postulate
- The famous *Michelson-Morley experiment* gave further support to the postulate that detecting an absolute reference velocity was not achievable. It should be stated here that, contrary to many alternative claims, it said little about the invariance of the speed of light with respect to the source and observer's velocity, as both source and observer were travelling together at the same velocity at all times.

A number of experiments have been conducted to test special relativity against rival theories. These include:

- *Kaufmann-Bucherer-Neumann experiments*—electron deflection in approximate agreement with Lorentz-Einstein prediction.
- *Fizeau experiment*—speed of light in moving media in accordance with relativistic velocity addition.
- *Kennedy-Thorndike experiment*—time dilation in accordance with Lorentz transformations.

- *Rossi-Hall experiment*—relativistic effects on a fast-moving particle's half-life.
- Experiments to test *emitter theory* demonstrated that the speed of light is independent of the speed of the emitter.
- *Hammar experiment*—no "aether flow obstruction".

In addition, particle accelerators routinely accelerate and measure the properties of particles moving at near the speed of light, where their behavior is completely consistent with relativity theory and inconsistent with the earlier Newtonian mechanics. These machines would simply not work if they were not engineered according to relativistic principles.

CHAPTER 4

PHOTOCHEMICAL REACTION

A photochemical reaction is a chemical reaction which is induced by light. Examples of photochemical organic reactions are electrocyclic reactions, photoisomerization and Norrish reactions.

The basic requirements for photochemical reactions are:

- the energy of the light source must correspond to an electronic transition between orbitals.
- the emitted light must be able to reach the targeted functional group without being blocked by the reactor, medium or other functional groups present.

Photoexcitation is the first step in a photochemical process where the reactant is elevated to an excited state of higher energy. Photosensitizers absorb radiation and transfer energy to the reactant. The opposite process is called quenching when a photoexited state is deactivated by a chemical reagent.

The first ever photochemical reaction was described by Trommsdorf in 1834 He observed that crystals of the compound a-santonin when exposed to sunlight turned yellow and burst. In a 2007 study the reaction was described as a succession of three steps taking place within a single crystal intermediate 2, the second one a dimerization in a Diels-Alder reaction (3) and the third one a intramolecular [2+2] cycloaddition (4). The bursting effect is attributed to a large change in crystal volume on dimerization.

Most photochemical transformations occur through a series of simple steps known as primary photochemical processes. One common example of these processes is the Excited State Proton Transfer (ESPT).

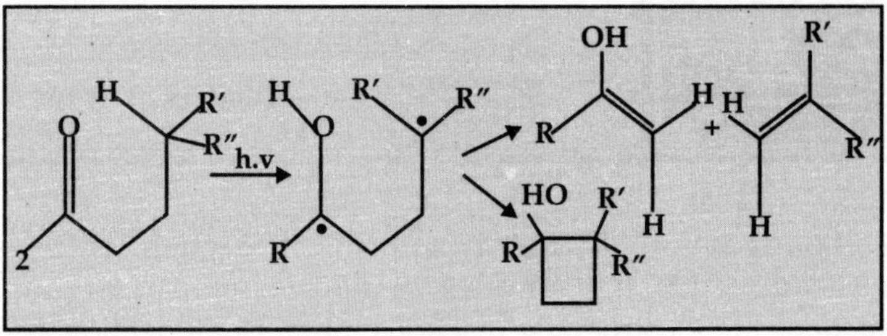

Fig. 4.1

Photochromism

Photochromism is the reversible transformation of a chemical species between two forms by the absorption of electromagnetic radiation, where the two forms have different absorption spectra. Trivially, this can be described as a reversible change of color upon exposure to light. The phenomenon was discovered in the late 1880s, including work by Markwald, who studied the reversible change of color of 2, 3, 4, 4-tetrachloronaphthalen-1(4H)-one in the solid state. He labeled this phenomenon "phototropy", and this name was used until the 1950s when Yehuda Hirshberg, of the Weizmann Institute of Science in Israel proposed the term "photochromism" Photochromism can take place in both organic and inorganic compounds, and also has its place in biological systems (for example retinal in the vision process).

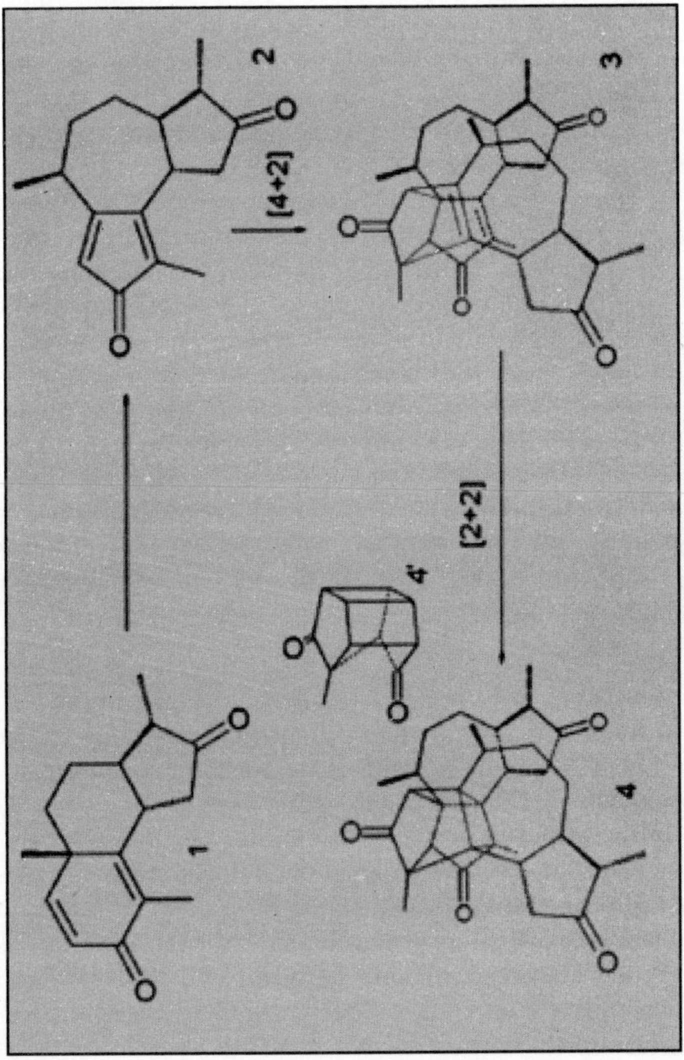

1

2

[4+2]

3

[2+2]

4'

4

Fig. 4.2

Overview

Photochromism does not have a rigorous definition, but is usually used to describe compounds that undergo a reversible photochemical reaction where an absorption band in the visible part of the electromagnetic spectrum changes dramatically in strength or wavelength. In many cases, an absorbance band is present in only one form. The degree of change required for a photochemical reaction to be dubbed "photochromic" is that which appears dramatic by eye, but in essence there is no dividing line between photochromic reactions and other photochemistry. Therefore, while the trans-cis isomerization of azobenzene is considered a photochromic reaction, the analogous reaction of stilbene is not. Since photochromism is just a special case of a photochemical reaction, almost any photochemical reaction type may be used to produce photochromism with appropriate molecular design. Some of the most common processes involved in photochromism are pericyclic reactions, cis-trans isomerizations, intramolecular hydrogen transfer, intramolecular group transfers, dissociation processes and electron transfers (oxidation-reduction).

Another somewhat arbitrary requirement of photochromism is that it requires the two states of the molecule to be thermally stable under ambient conditions for a reasonable time. All the same, nitrospiropyran (which back-isomerizes in the dark over ~10 minutes at room temperature) is considered photochromic. All photochromic molecules back-isomerize to their more stable form at some rate, and this back-isomerization is accelerated by heating. There is therefore a close relationship between photochromic and thermochromic compounds. The timescale of thermal back-isomerization is important for applications, and may be molecularly engineered. Photochromic compounds considered to be "thermally stable" include some diarylethenes, which do not back isomerize even after heating at 80° C for 3 months.

Since photochromic chromophores are dyes, and operate according to well-known reactions, their molecular engineering to fine-tune their properties can be achieved relatively easily using known design models, quantum mechanics calculations, and experimentation. In particular, the tuning of absorbance bands to particular parts of the spectrum and the engineering of thermal stability have received much attention.

Sometimes and particularly in the dye industry, the term "irreversible photochromic" is used to describe materials that undergo a permanent color change upon exposure to ultraviolet or visible light radiation. Because by definition photochromics are reversible, there is technically no such thing as an "irreversible photochromic"—this is loose usage, and these compounds are better referred to as "photochangable" or "photoreactive" dyes.

Apart from the qualities already mentioned, several other properties of photochromics are important for their use. These include:

- *Quantum yield* of the photochemical reaction. This determined the efficiency of the photochromic change with respect to the amount of light absorbed. The quantum yield of isomerization can be strongly dependent on conditions.
- *Fatigue resistance*. In photochromic materials, fatigue refers to the loss of reversibility by processes such as photodegradation, photobleaching, photooxidation, and other side reactions. All photochromics suffer fatigue to some extent, and its rate is strongly dependent on the activating light and the conditions of the sample.
- *Photostationary state*. Photochromic materials have two states, and their interconversion can be controlled using different wavelengths of light. Excitation with any given wavelength of light will result in a mixture of the two states at a particular ratio, called the "photostationary state". In a perfect system, there would exist wavelengths that can be used to provide 1:0 and 0:1

ratios of the isomers, but in real systems this is not possible, since the active absorbance bands always overlap to some extent.

- *Polarity and solubility*. In order to incorporate photochromics in working systems, they suffer the same issues as other dyes. They are often charged in one or more state, leading to very high polarity and possible large changes in polarity. They also often contain large conjugated systems that limit their solubility.

Photochromic Complexes

A photochromic complex is a kind of chemical compound that has photoresponsive parts on its ligand. These complexes have a specific structure: photoswitchable organic compounds are attached to metal complexes. For the photocontrollable parts, thermally and photochemically stable chromophores (azobenzene, diarylethene, spiropyran, etc.) are usually used. And for the metal complexes, a wide variety of compounds that have various functions (redox response, luminescence, magnetism, etc.) are applied.

The photochromic parts and metal parts are so close that they can affect each other's molecular orbitals. The physical properties of these compounds shown by parts of them (i.e., chromophores or metals) thus can be controlled by switching their other sites by external stimuli. For example, photoisomerization behaviors of some complexes can be switched by oxidation and reduction of their metal parts. Some other compounds can be changed in their luminescence behavior, magnetic interaction of metal sites, or stability of metal-to-ligand coordination by photoisomerization of their photochromic parts.

Classes of Photochromic Materials

Photochromic molecules can belong to various classes: triarylmethanes, stilbenes, azastilbenes, nitrones, fulgides, spiropyrans, naphthopyrans, spiro-oxazines, quinones and others.

Spiropyrans and Sprioxazines

One of the oldest, and perhaps the most studied, families of photochromes are the spiropyrans. Very closely related to these are the spirooxazines. For example, the spiro form of an oxazine is a colorless leuco dye; the conjugated system of the oxazine and another aromatic part of the molecule is separated by a sp^3-hybridized "spiro" carbon. After irradiation with UV light, the bond between the spiro-carbon and the oxazine breaks, the ring opens, the spiro carbon achieves sp^2 hybridization and becomes planar, the aromatic group rotates, aligns its p-orbitals with the rest of the molecule, and a conjugated system forms with ability to absorb photons of visible light, and therefore appear colorful. When the UV source is removed, the molecules gradually relax to their ground state, the carbon-oxygen bond reforms, the spiro-carbon becomes sp^3 hybridized again, and the molecule returns to its colorless state.

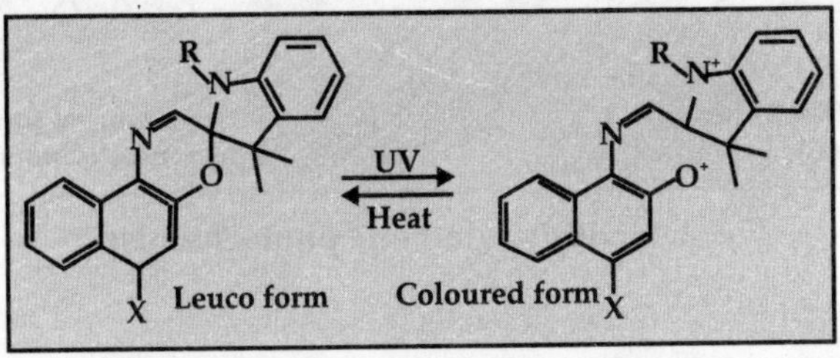

Fig. 4.3: Spiro-mero photochromism

This class of photochromes in particular are thermodynamically unstable in one form and revert to the stable form in the dark unless cooled to low temperatures. Their lifetime can also be affected by exposure to UV light. Like most organic dyes they are susceptible to degradation by oxygen and free radicals. Incorporation of the dyes into a polymer matrix, adding a stabilizer, or providing a barrier to oxygen and chemicals by other means prolongs their lifetime.

Diarylethenes

The "diarylethenes" were first introduced by Irie and have since gained widespread interest, largely on account of their high thermodynamic stability. They operate by means of a 6-pi electrocyclic reaction, the thermal analog of which is impossible due to steric hindrance. Pure photochromic dyes usually have the appearance of a crystalline powder, and in order to achieve the color change, they usually have to be dissolved in a solvent or dispersed in a suitable matrix. However, some diarylethenes have so little shape change upon isomerization that they can be converted while remaining in crystalline form.

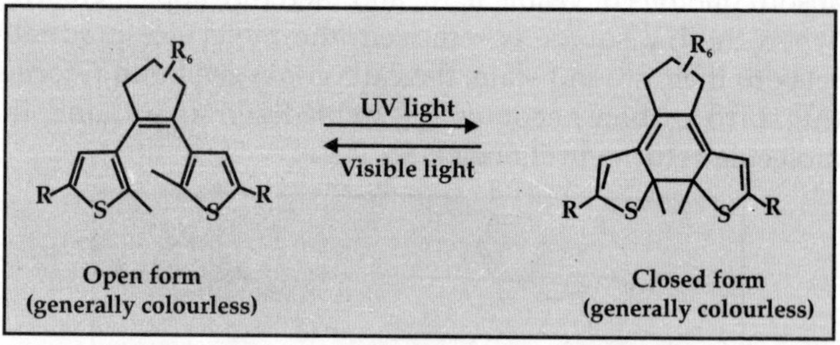

Fig. 4.4: Dithienylethene photochemistry

Azobenzenes

The photochromic trans-cis isomerization of azobenzenes has been used extensively in molecular switches, often taking advantage of its shape change upon isomerization to produce a supramolecular result. In particular, azobenzenes incorporated into crown ethers give switchable receptors and azobenzenes in monolayers can provide light-controlled changes in surface properties.

Photochromic Quinones

Some quinones, and phenoxynaphthacene quinone in particular, have photochromicity resulting from the ability of the phenyl group to migrate from one oxygen atome to another. Quinones with good thermal stability have been

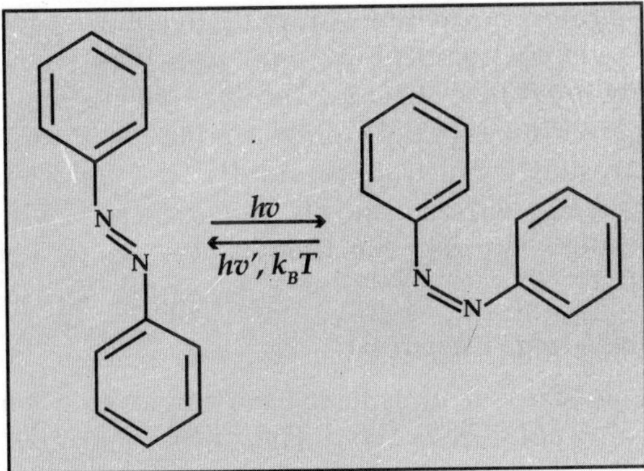

Fig. 4.5: Azobenzene photoisomerization

prepared, and they also have the additional feature of redox activity, leading to the construction of many-state molecular switches that operate by a mixture of photonic and electronic stimuli.

Inorganic Photochromics

Many inorganic substances also exhibit photochromic properties, often with much better resistance to fatigue than organic photochromics. In particular, silver chloride is extensively used in the manufacture of photochromic lenses. Other silver and zinc halides are also photochromic.

Applications

Sunglasses

One of the most famous reversible photochromic applications is color changing lenses for sunglasses, as found in eye-glasses. The largest limitation in using PC technology is that the materials cannot be made stable enough to withstand thousands of hours of outdoor exposure so long-term outdoor applications are not appropriate at this time.

The switching speed of photochromic dyes is highly sensitive to the rigidity of the environment around the dye.

As result, they switch most rapidly in solution and slowest in the rigid environment like a polymer lens. Recently it has been reported that attaching flexible, low Tg polymers [for example siloxanes or poly (butyl acrylate)] to the dyes allows them to switch much more rapidly in a rigid lens Some spirooxazines with siloxane polymers attached switch at near solution-like speeds even though they are in a rigid lens matrix.

Supramolecular Chemistry

Photochromic units have been employed extensively in supramolecular chemistry. Their ability to give a light-controlled reversible shape change means that they can be used to make or break molecular recognition motifs, or to cause a consequent shape change in their surroundings. Thus, photochromic units have been demonstrated as components of molecular switches. The coupling of photochromic units to enzymes or enzyme cofactors even provides the ability to reversibly turn enzymes "on" and "off", by altering their shape or orientation in such a way that their functions are either "working" or "broken".

Data Storage

The possibility of using photochromic compounds for data storage was first suggested in 1956 by Yehuda Hirshberg Since that time, there have been many investigations by various academic and commercial groups, particularly in the area of 3D optical data storage which promises discs that can hold a terabyte of data. Initially, issues with thermal back-reactions and destructive reading dogged these studies, but more recently more-stable.

Novelty Items

Reversible photochromics are also found in applications such as toys, cosmetics, clothing and industrial applications. If necessary, they can be made to change between desired colors by combination with a permanent pigment.

CHAPTER 5 LIGHT

Light, or visible light, is electromagnetic radiation of a wavelength that is visible to the human eye (about 400-700 nm), or up to 380-750 nm. In the broader field of physics, light is sometimes used to refer to electromagnetic radiation of all wavelengths, whether visible or not.

Three primary properties of light are:

1. Intensity;
2. Frequency or wavelength; and
3. Polarization.

Light can exhibit properties of both waves and particles (photons). This property is referred to as wave–particle duality. The study of light, known as optics, is an important research area in modern physics.

Speed of Light

The speed of light in a vacuum is presently defined to be exactly 299,792,458 m/s (about 186,282.397 miles per second). This definition of the speed of light means that the metre is now defined in terms of the speed of light. Light always travels at a constant speed, even between particles of a substance through which it is shining. Photons excite the adjoining particles that in turn transfer the energy to the neighbour. This may appear to slow the beam down though its trajectory in realtime. The time lost between entry and exit accounts to the displacement of energy through the substance between each particle that is excited.

Different physicists have attempted to measure the speed of light throughout history. Galileo attempted to measure the speed of light in the seventeenth century. An early experiment to measure the speed of light was conducted by Ole Rømer, a Danish physicist, in 1676. Using a telescope, Ole observed the motions of Jupiter and one of its moons, Io. Noting discrepancies in the apparent period of Io's orbit, Rømer calculated that light takes about 22 minutes to traverse the diameter of Earth's orbit. Unfortunately, its size was not known at that time. If Ole had known the diameter of the Earth's orbit, he would have calculated a speed of 227,000,000 m/s.

Another, more accurate, measurement of the speed of light was performed in Europe by Hippolyte Fizeau in 1849. Fizeau directed a beam of light at a mirror several kilometers away. A rotating cog wheel was placed in the path of the light beam as it traveled from the source, to the mirror and then returned to its origin. Fizeau found that at a certain rate of rotation, the beam would pass through one gap in the wheel on the way out and the next gap on the way back. Knowing the distance to the mirror, the number of teeth on the wheel, and the rate of rotation, Fizeau was able to calculate the speed of light as 313,000,000 m/s.

Léon Foucault used an experiment which used rotating mirrors to obtain a value of 298,000,000 m/s in 1862. Albert A. Michelson conducted experiments on the speed of light from 1877 until his death in 1931. He refined Foucault's methods in 1926 using improved rotating mirrors to measure the time it took light to make a round trip from Mt. Wilson to Mt. San Antonio in California. The precise measurements yielded a speed of 299,796,000 m/s.

Two independent teams of physicists were able to bring light to a complete standstill by passing it through a Bose-Einstein Condensate of the element rubidium, one led by Dr. Lene Vestergaard Hau of Harvard University and the Rowland Institute for Science in Cambridge, Mass., and the other by Dr. Ronald L. Walsworth and Dr. Mikhail D. Lukin of the Harvard-Smithsonian Center for Astrophysics, also in Cambridge.

Refraction

When a beam of light crosses the boundary between a vacuum and another medium, or between two different media, the wavelength of the light changes, but the frequency remains constant. If the beam of light is not orthogonal (or rather normal) to the boundary, the change in wavelength results in a change in the direction of the beam. This change of direction is known as refraction.

The refractive quality of lenses is frequently used to manipulate light in order to change the apparent size of images. Magnifying glasses, spectacles, contact lenses, microscopes and refracting telescopes are all examples of this manipulation.

Light refraction is the main basis of measurement for gloss. Gloss is measured using a glossmeter, and an objects refractive index is what the glossmeter analyses.

Optics

The study of light and the interaction of light and matter is termed optics. The observation and study of optical phenomena such as rainbows and the aurora borealis offer many clues as to the nature of light as well as much enjoyment.

Light Sources

There are many sources of light. The most common light sources are thermal: a body at a given temperature emits a characteristic spectrum of black-body radiation. Examples include sunlight (the radiation emitted by the chromosphere of the Sun at around 6,000 K peaks in the visible region of the electromagnetic spectrum), incandescent light bulbs (which emit only around 10% of their energy as visible light and the remainder as infrared), and glowing solid particles in flames. The peak of the blackbody spectrum is in the infrared for relatively cool objects like human beings. As the temperature increases, the peak shifts to shorter wavelengths, producing first a red glow, then a white one, and finally a blue color as the peak moves out of the visible part of the spectrum and into the ultraviolet. These colors can be seen when metal is heated to "red hot" or "white hot". Blue thermal emission is not often seen. The commonly seen blue colour in a gas flame or a welder's torch is in fact due to molecular emission, notably by CH radicals (emitting a wavelength band around 425 nm).

Atoms emit and absorb light at characteristic energies. This produces "emission lines" in the spectrum of each atom. Emission can be spontaneous, as in light-emitting diodes, gas discharge lamps (such as neon lamps and neon signs, mercury-vapor lamps, etc.), and flames (light from the hot gas itself—so, for example, sodium in a gas flame emits characteristic yellow light). Emission can also be stimulated, as in a laser or a microwave maser.

Deceleration of a free charged particle, such as an electron, can produce visible radiation: cyclotron radiation, synchrotron radiation, and bremsstrahlung radiation are all examples of this. Particles moving through a medium faster than the speed of light in that medium can produce visible Cherenkov radiation.

Certain chemicals produce visible radiation by chemoluminescence. In living things, this process is called bioluminescence. For example, fireflies produce light by this means, and boats moving through water can disturb plankton which produce a glowing wake.

Certain substances produce light when they are illuminated by more energetic radiation, a process known as fluorescence. This is used in fluorescent lights. Some substances emit light slowly after excitation by more energetic radiation. This is known as phosphorescence.

Phosphorescent materials can also be excited by bombarding them with subatomic particles. Cathodoluminescence is one example of this. This mechanism is used in cathode ray tube televisions.

Certain other mechanisms can produce light:

- scintillation
- electroluminescence
- sonoluminescence
- triboluminescence
- Cherenkov radiation

When the concept of light is intended to include very-high-energy photons (gamma rays), additional generation mechanisms include:

- radioactive decay
- particle–antiparticle annihilation

Theories about Light

Hindu and Buddhist Theories

In ancient India, the Hindu schools of *Samkhya* and *Vaisheshika*, from around the 6th–5th century BC, developed theories on light. According to the *Samkhya* school, light is one of the five fundamental "subtle" elements (*tanmatra*) out of which emerge the gross elements. The atomicity of these elements is not specifically mentioned and it appears that they were actually taken to be continuous.

On the other hand, the *Vaisheshika* school gives an atomic theory of the physical world on the non-atomic ground of ether, space and time. The basic atoms are those of earth (*prthivi*), water (*apas*), fire (*tejas*), and air (*vayu*), that should not be confused with the ordinary meaning of these terms. These atoms are taken to form binary molecules that combine further to form larger molecules. Motion is defined in terms of the movement of the physical atoms and it appears that it is taken to be non-instantaneous. Light rays are taken to be a stream of high velocity of *tejas* (fire) atoms. The particles of light can exhibit different characteristics depending on the speed and the arrangements of the *tejas* atoms. Around the first century BC, the *Vishnu Purana* refers to sunlight as the "the seven rays of the sun".

Later in 499, Aryabhatta, who proposed a heliocentric solar system of gravitation in his *Aryabhatiya*, wrote that the planets and the Moon do not have their own light but reflect the light of the Sun.

The Indian Buddhists, such as Dignaga in the 5th century and Dharmakirti in the 7th century, developed a type of atomism that is a philosophy about reality being composed of atomic entities that are momentary flashes of light or energy. They viewed light as being an atomic entity equivalent to energy, similar to the modern concept of photons, though they also viewed all matter as being composed of these light/energy particles.

> Mixing the three colours, have produced all the objects of sight.
>
> —*Rigveda*

Greek and Thellenistic Theories

In the fifth century BC, Empedocles postulated that everything was composed of four elements; fire, air, earth and water. He believed that Aphrodite made the human eye out of the four elements and that she lit the fire in the eye which shone out from the eye making sight possible. If this were true, then one could see during the night just as well as during the day, so Empedocles postulated an interaction between rays from the eyes and rays from a source such as the sun.

In about 300 BC, Euclid wrote Optica, in which he studied the properties of light. Euclid postulated that light travelled in straight lines and he described the laws of reflection and studied them mathematically. He questioned that sight is the result of a beam from the eye, for he asks how one sees the stars immediately, if one closes one's eyes, then opens them at night. Of course if the beam from the eye travels infinitely fast this is not a problem.

In 55 BC, *Lucretius*, a Roman who carried on the ideas of earlier Greek atomists, wrote:

> "The light and heat of the sun; these are composed of minute atoms which, when they are shoved off, lose no time in shooting right across the interspace of air in the direction imparted by the shove."
>
> —*On the nature of the Universe*

Despite being similar to later particle theories, Lucretius's views were not generally accepted and light was still theorized as emanating from the eye.

Ptolemy (c. 2nd century) wrote about the refraction of light in his book Optics, and developed a theory of vision whereby objects are seen by rays of light emanating from the eyes.

Optical Theory

The Muslim scientist, Ibn al-Haytham (965-1040), known as Alhacen or Alhazen in the West, in his Book of Optics (1021), developed a broad theory that explained vision, using geometry and anatomy, which stated that each point on an illuminated area or object radiates light rays in every direction, but that only one ray from each point, which strikes the eye perpendicularly, can be seen. The other rays strike at different angles and are not seen. He described the pinhole camera and invented the camera obscura, which produces an inverted image, and used it as an example to support his argument. This contradicted Ptolemy's theory of vision that objects are seen by rays of light emanating from the eyes. Alhacen held light rays to be streams of minute energy particles[that travelled at a finite speed. He improved Ptolemy's theory of the refraction of light, and went on to discover the laws of refraction.

He also carried out the first experiments on the dispersion of light into its constituent colors. His major work *Kitab al-Manazir* (*Book of Optics*) was translated into Latin in the Middle Ages, as well his book dealing with the colors of sunset. He dealt at length with the theory of various physical phenomena like shadows, eclipses, the rainbow. He also attempted to explain binocular vision, and gave a correct explanation of the apparent increase in size of the sun and the moon when near the horizon, known as the moon illusion. Because of his extensive experimental research on optics, Ibn al-Haytham is considered the "father of modern optics"

Ibn al-Haytham also correctly argued that we see objects because the sun's rays of light, which he believed to be streams of tiny energy particles travelling in straight lines, are reflected from objects into our eyes. He understood that light must travel at a large but finite velocity, and that refraction is caused by the velocity being different in different substances. He also studied spherical and parabolic mirrors, and understood how refraction by a lens will allow images to be focused and magnification to take place. He understood mathematically why a spherical mirror produces aberration.

Avicenna (980-1037) agreed that the speed of light is finite, as he "observed that if the perception of light is due to the emission of some sort of particles by a luminous source, the speed of light must be finite." Abu Rayhan al-Biruni (973-1048) also agreed that light has a finite speed, and he was the first to discover that the speed of light is much faster than the speed of sound In the late 13th and early 14th centuries, Qutb al-Din al-Shirazi (1236-1311) and his student Kamal al-Din al-Farisi (1260-1320) continued the work of Ibn al-Haytham, and they were the first to give the correct explanations for the rainbow phenomenon.

The 'Plenum'

René Descartes (1596-1650) held that light was a disturbance of the plenum, the continuous substance of which the universe was composed. In 1637 he published a theory of the refraction of light that assumed, incorrectly, that light travelled faster in a denser medium than in a less dense medium. Descartes arrived at this conclusion by analogy with the behaviour of sound waves. Although Descartes was incorrect about the relative speeds, he was correct in assuming that light behaved like a wave and in concluding that refraction could be explained by the speed of light in different media. As a result, Descartes' theory is often regarded as the forerunner of the wave theory of light.

Particle Theory

Pierre Gassendi (1592-1655), an atomist, proposed a particle theory of light which was published posthumously in the 1660s. Isaac Newton studied Gassendi's work at an early age, and preferred his view to Descartes' theory of the plenum. He stated in his Hypothesis of Light of 1675 that light was composed of corpuscles (particles of matter) which were emitted in all directions from a source. One of Newton's arguments against the wave nature of light was that waves were known to bend around obstacles, while light travelled only in straight lines. He did, however, explain the phenomenon of the diffraction of light (which had been observed by Francesco Grimaldi) by allowing that a light particle could create a localised wave in the aether.

Newton's theory could be used to predict the reflection of light, but could only explain refraction by incorrectly assuming that light accelerated upon entering a denser medium because the gravitational pull was greater. Newton published the final version of his theory in his Opticks of 1704. His reputation helped the particle theory of light to hold sway during the 18th century. The particle theory of light led Laplace to argue that a body could be so massive that light could not escape from it. In other words it would become what is now called a black hole. Laplace withdrew his suggestion when the wave theory of light was firmly established. A translation of his essay appears in The large scale structure of space-time, by Stephen Hawking and George F. R. Ellis.

Wave Theory

In the 1660s, Robert Hooke published a wave theory of light. Christiaan Huygens worked out his own wave theory of light in 1678, and published it in his Treatise on light in 1690. He proposed that light was emitted in all directions as a series of waves in a medium called the Luminiferous ether. As waves are not affected by gravity, it was assumed that they slowed down upon entering a denser medium.

The wave theory predicted that light waves could interfere with each other like sound waves (as noted around 1800 by Thomas Young), and that light could be polarized. Young showed by means of a diffraction experiment that light behaved as waves. He also proposed that different colors were caused by different wavelengths of light, and explained color vision in terms of three-colored receptors in the eye.

Another supporter of the wave theory was Leonhard Euler. He argued in Nova theoria lucis et colorum (1746) that diffraction could more easily be explained by a wave theory.

Later, Augustin-Jean Fresnel independently worked out his own wave theory of light, and presented it to the Académie des Sciences in 1817. Simeon Denis Poisson added to Fresnel's mathematical work to produce a convincing argument in favour of the wave theory, helping to overturn Newton's corpuscular theory.

The weakness of the wave theory was that light waves, like sound waves, would need a medium for transmission. A hypothetical substance called the luminiferous aether was proposed, but its existence was cast into strong doubt in the late nineteenth century by the Michelson-Morley experiment.

Newton's corpuscular theory implied that light would travel faster in a denser medium, while the wave theory of Huygens and others implied the opposite. At that time, the speed of light could not be measured accurately enough to decide which theory was correct. The first to make a sufficiently accurate measurement was Léon Foucault, in 1850. His result supported the wave theory, and the classical particle theory was finally abandoned.

Electromagnetic Theory

In 1845, Michael Faraday discovered that the plane of polarization of linearly polarized light is rotated when the light rays travel along the magnetic field direction in the presence of a transparent dielectric, an effect now known as Faraday rotation. This was the first evidence that light was related to electromagnetism. In 1846 he speculated that light might be some form of disturbance propagating along magnetic field lines. Faraday proposed in 1847 that light was a high-frequency electromagnetic vibration, which could propagate even in the absence of a medium such as the ether.

Faraday's work inspired James Clerk Maxwell to study electromagnetic radiation and light. Maxwell discovered that self-propagating electromagnetic waves would travel through space at a constant speed, which happened to be equal to the previously measured speed of light. From this, Maxwell concluded that light was a form of electromagnetic radiation: he first stated this result in 1862 in On Physical Lines of Force. In 1873, he published A Treatise on Electricity and Magnetism, which contained a full mathematical description of the behaviour of electric and magnetic fields, still known as Maxwell's equations. Soon after, Heinrich Hertz confirmed Maxwell's theory experimentally by generating and detecting radio waves in the laboratory, and demonstrating that these waves behaved exactly like visible light, exhibiting properties such as reflection, refraction, diffraction, and interference. Maxwell's theory and Hertz's experiments led directly to the development of modern radio, radar, television, electromagnetic imaging, and wireless communications.

The Special Theory of Relativity

The wave theory was wildly successful in explaining nearly all optical and electromagnetic phenomena, and was a great triumph of nineteenth century physics. By the late nineteenth century, however, a handful of experimental anomalies remained that could not be explained by or were in direct conflict with the wave theory. One of these anomalies involved a controversy over the speed of light. The constant speed of light predicted by Maxwell's equations and confirmed by the Michelson-Morley experiment contradicted the mechanical laws of motion that had been unchallenged since the time of Galileo, which stated that all speeds were relative to the speed of the observer. In 1905, Albert Einstein resolved this paradox by revising the Galilean model of space and time to account for the constancy of the speed of light. Einstein formulated his ideas in his special theory of relativity, which radically altered humankind's understanding of space and time. Einstein also demonstrated a previously unknown fundamental equivalence between energy and mass with his famous equation

$$E = mc^2$$

where E is energy, m is rest mass, and c is the speed of light in a vacuum.

Particle Theory Revisited

Another experimental anomaly was the photoelectric effect, by which light striking a metal surface ejected electrons from the surface, causing an electric current to flow across an applied voltage. Experimental measurements demonstrated that the energy of individual ejected electrons was proportional to the frequency, rather than the intensity, of the light. Furthermore, below a certain minimum frequency, which depended on the particular metal, no current would flow regardless of the intensity. These observations clearly contradicted the wave theory, and for years physicists tried in vain to find an explanation. In 1905, Einstein solved this puzzle as well, this time by resurrecting the particle theory of light to explain the observed effect. Because of the preponderance of evidence in favor of the wave theory, however, Einstein's ideas were met initially by great skepticism among established physicists. But eventually Einstein's explanation of the photoelectric effect would triumph, and it ultimately formed the basis for wave–particle duality and much of quantum mechanics.

Quantum Theory

A third anomaly that arose in the late 19th century involved a contradiction between the wave theory of light and measurements of the electromagnetic spectrum emitted by thermal radiators, or so-called black bodies. Physicists struggled with this problem, which later became known as the ultraviolet catastrophe, unsuccessfully for many years. In 1900, Max Planck developed a new theory of black-body radiation that explained the observed spectrum correctly. Planck's theory was based on the idea that black bodies emit light (and other electromagnetic radiation) only as discrete bundles or packets of energy. These packets were called quanta, and the particle of light was given the name photon, to correspond with other particles being described around this time, such as the electron and proton. A photon has an energy, E, proportional to its frequency, f, by

$$E = hf = \frac{hc}{\lambda}$$

where h is Planck's constant, λ is the wavelength and c is the speed of light. Likewise, the momentum p of a photon is also proportional to its frequency and inversely proportional to its wavelength:

$$p = \frac{E}{c} = \frac{hf}{c} = \frac{h}{\lambda}$$

As it originally stood, this theory did not explain the simultaneous wave- and particle-like natures of light, though Planck would later work on theories that did. In 1918, Planck received the Nobel Prize in Physics for his part in the founding of quantum theory.

Wave-particle Duality

The modern theory that explains the nature of light includes the notion of wave–particle duality, described by Albert Einstein in the early 1900s, based on his study of the photoelectric effect and Planck's results. Einstein asserted that the energy of a photon is proportional to its frequency. More generally, the theory states that everything has both a particle nature and a wave nature, and various experiments can be done to bring out one or the other. The particle nature is more easily discerned if an object has a large mass, and it was not until a bold proposition by Louis de Broglie in 1924 that the scientific community realized that electrons also exhibited wave–particle duality. The wave nature of electrons was experimentally demonstrated by Davission and Germer in 1927. Einstein received the Nobel Prize in 1921 for his work with the wave–particle duality on photons (especially explaining the photoelectric effect thereby), and de Broglie followed in 1929 for his extension to other particles.

Quantum Electrodynamics

The quantum mechanical theory of light and electromagnetic radiation continued to evolve through the 1920s and 1930s, and culminated with the development during the 1940s of the theory of quantum electrodynamics, or QED. This so-called quantum field theory is among the most comprehensive and experimentally successful theories ever formulated to explain a set of natural phenomena. QED was developed primarily by physicists Richard Feynman, Freeman Dyson, Julian Schwinger, and Shin-Ichiro Tomonaga. Feynman, Schwinger, and Tomonaga shared the 1965 Nobel Prize in Physics for their contributions.

Light Pressure

Radiation Pressure

Light pushes on objects in its path, just as the wind would do. This pressure is most easily explainable in particle theory: photons hit and transfer their momentum. Light pressure can cause asteroids to spin faster acting on their irregular shapes as on the vanes of a windmill. The possibility to make solar sails that would accelerate spaceships in space is also under investigation.

Although the motion of the Crookes radiometer was originally attributed to light pressure, this interpretation is incorrect; the characteristic Crookes rotation is the result of a partial vacuum This should not be confused with the Nichols radiometer, in which the motion is directly caused by light pressure.

Spirituality

The sensory perception of light plays a central role in spirituality (vision, enlightenment, darshan, Tabor Light), and the presence of light as opposed to its absence (darkness) is a common Western metaphor of good and evil, knowledge and ignorance, and similar concepts.

POLARIZATION

Polarization (Brit. polarisation) is a property of waves that describes the orientation of their oscillations. For transverse waves such as many electromagnetic waves, it describes the orientation of the oscillations in the plane perpendicular to the wave's direction of travel. The oscillations may be oriented in a single direction (linear polarization), or the oscillation direction may rotate as the wave travels (circular or elliptical polarization). For longitudinal waves such as sound waves in fluids the direction of oscillation is by definition along the direction of travel (i.e. polarization is not possible). Guided modes in waveguides and optical fibers can carry waves with both transverse and longitudinal oscillations. Such waves do have polarization.

Polarization is used in areas of science and technology dealing with wave propagation, such as optics, seismology, and telecommunications. For electromagnetic waves such as light, the polarization is described by specifying the direction of the wave's electric field. According to the Maxwell equations, the direction of the magnetic field is uniquely determined for a specific electric field distribution and polarization. For transverse sound waves in a solid, the polarization is associated with the direction of the shear stress in the plane perpendicular to the propagation direction.

Basics: Plane Waves

The simplest manifestation of polarization to visualize is that of a plane wave, which is a good approximation of most light waves (a plane wave is a wave with infinitely long and wide wavefronts). For plane waves the transverse condition requires that the electric and magnetic field be perpendicular to the direction of propagation and to each other. Conventionally, when considering polarization, the electric field vector is described and the magnetic field is ignored since it is perpendicular to the electric field and proportional to it. The electric field vector of a plane wave may be arbitrarily divided into two perpendicular components labeled x and y (with z indicating the direction of travel). For a simple harmonic wave, where the amplitude of the electric vector varies in a sinusoidal manner in time, the two components have exactly the same frequency. However, these components have two other defining characteristics that can differ. First, the two components may not have the same amplitude. Second, the two components may not have the same phase, that is they may not reach their maxima and minima at the same time. Mathematically, the electric field of a plane wave can be written as,

$$\vec{E}(\vec{r}, t) = \text{Re}\,[(A_x, A_y.e^{i\phi}, 0)e^{i(kz-\omega t)}]$$

or alternatively,

$$\vec{E}(r, \vec{t}) = (Ax.\cos(kz - \omega t), Ay.\cos(kz - \omega t + \phi), 0)$$

where A_x and A_y are the amplitudes of the x and y directions and f is the relative phase between the two components.

The "Cartesian" decomposition of the electric field into x and y components is, of course, arbitrary. Plane waves of any polarization can be described instead by combining waves of opposite circular polarization, for example. The Cartesian polarization decomposition is natural when dealing with reflection from surfaces, birefringent materials, or synchrotron radiation. The circularly polarized modes are a more useful basis for the study of light propagation in stereoisomers.

Incoherent Radiation

In nature, electromagnetic radiation is often produced by a large number of individual sources, producing waves independently of each other. This type of light is described as incoherent. In general there is no single frequency but rather a spectrum of different frequencies present, and even if filtered to an arbitrarily narrow frequency range, there may not be a consistent state of polarization. However, this does not mean that polarization is only a feature of coherent radiation. Incoherent radiation may show statistical correlation between the components of the electric field, which can be interpreted as partial polarization. In general it is possible to describe an observed wave field as the sum of a completely incoherent part (no correlations) and a completely polarized part. One may then describe the light in terms of the degree of polarization, and the parameters of the polarization ellipse.

Parameterizing Polarization

Full information on a completely polarized state is also provided by the amplitude and phase of oscillations in two components of the electric field vector in the plane of polarization. This representation was used above to show how different states of polarization re possible. The amplitude and phase information can be coeniently represented as a two-dimensional complex vector (the Jones vector):

$$e = \begin{bmatrix} a_1 e^{i\theta_1} \\ a_2 e^{i\theta_2} \end{bmatrix}$$

Here a_1 and a_2 denote the amplitude of the wave in the two compnents of the electric field vector, while θ_1 and θ_2 represent he phases. The product of a Jones vector with a complex number of unit modulus gives a different Jones vector representing the same ellipse, and thus the same state of polarization. The physal electric field, as the real part of the Jones vector, would be ltered but the polarization state itself is independent of absolute pase. The basis vectors used to represent the Jones vector need not represent linear polarization states (i.e. be real). In generalany two orthogonal states can be used, where an orthogonal vector pair is formally defined as one having a zero inner product. A common choice is left and right circular polarizations, for example to model the different propagation of waves in two such components in circularly birefringent media (see below) or signal paths of coherent detectors sensitive to circular polarization.

Regardless of whether polarization ellipses are represented using geometric parameters or Jones vectors, implicit in the parameterization is the orientation of the coordinate frame. This permits a degree of freedom, namely rotation about the propagation direction. When considering light that is propagating parallel to the surface of the Earth, the terms "horizontal" and "vertical" polarization are often used, with the former being associated with the first component of the Jones vector, or zero azimuth angle. On the other hand, in astronomy the equatorial coordinate system is generally used instead, with the zero azimuth (or position angle, as it is more commonly called in astronomy to avoid confusion with the horizontal coordinate system) corresponding to due north. Another coordinate system frequently used relates to the plane made by the propagation direction and a vector normal to the plane of a reflecting surface. This is known as the plane of incidence. The rays in this plane are illustrated in the diagram to the right. The component of the electric field parallel to this plane is termed p-like (parallel) and the component perpendicular to this plane is termed s-like (from senkrecht, German for perpendicular). Light with a p-like electric field is said to be p-polarized, pi-polarized, tangential plane polarized, or is said to be a transverse-magnetic (TM) wave. Light with an s-like electric field is s-polarized, also known as sigma-polarized or sagittal plane polarized, or it can be called a transverse-electric (TE) wave.

In the case of partially-polarized radiation, the Jones vector varies in time and space in a way that differs from the constant rate of phase rotation of monochromatic, purely-polarized waves. In this case, the wave field is likely stochastic, and only statistical information can be gathered about the variations and correlations between components of the electric field. This information is embodied in the coherency matrix:

$$\Psi = \langle ee^{+} \rangle$$

$$= \left\langle \begin{bmatrix} e1e^{*}_{1}, e1e^{*}_{2} \\ e2e^{*}_{1}, e2e^{*}_{2} \end{bmatrix} \right\rangle$$

$$= \left\langle \begin{bmatrix} a_1^2 & a_1 a_2 e^{-i(\theta 1-\theta 2)} \\ a_1 a_2 e^{-i(\theta 1-\theta 2)} & a_2^2 \end{bmatrix} \right\rangle$$

where angular brackets denote averaging over many wave cycles. Several variants of the coherency matrix have been proposed the Wiener coherency matrix and the spectral coherency matrix of Richard Barakat measure the coherence of a spectral decomposition of the signal, while the Wolf coherency matrix averages over all time/frequencies.

The coherency matrix contains all of the information on polarization that is obtainable using second order statistics. It can be decomposed into the sum of two idempotent matrices, corresponding to the eigenvectors of the coherency matrix, each representing a polarization state that is orthogonal to the other. An alternative decomposition is into completely polarized (zero determinant) and unpolarized (scaled identity matrix) components. In either case, the operation of summing the components corresponds to the incoherent superposition of waves from the two components. The latter case gives rise to the concept of the "degree of polarization"; i.e., the fraction of the total intensity contributed by the completely polarized component.

The coherency matrix is not easy to visualize, and it is therefore common to describe incoherent or partially polarized radiation in terms of its total intensity (I), (fractional) degree of polarization (p), and the shape parameters of the polarization ellipse. An alternative and mathematically convenient description is given by the Stokes parameters, introduced by George Gabriel Stokes in 1852. The relationship of the Stokes parameters to intensity and polarization ellipse parameters is shown in the equations and figure below.

$$S_0 = I$$

$$S_1 = Ip \cos 2\Psi \cos 2\chi$$

$$S_2 = Ip \sin 2\Psi \cos 2\chi$$

$$S_3 = Ip \sin 2\chi$$

Here Ip, 2χ and 2χ are the spherical coordinates of the polarization state in the three-dimensional space of the last three Stokes parameters. Note the factors of two before ψ and χ corresponding respectively to the facts that any polarization ellipse is indistinguishable from one rotated by 180°, or one with the semi-axis lengths swapped accompanied by a 90° rotation. The Stokes parameters are sometimes denoted I, Q, U and V.

The Stokes parameters contain all of the information of the coherency matrix, and are related to it linearly by means of the identity matrix plus the three Pauli matrices:

$$\Psi = \frac{1}{2} \sum_{j=0}^{3} S_j \sigma_j$$

where

$$\sigma_0 = \begin{bmatrix} 1 & 0 \\ 0 & 1 \end{bmatrix} \quad \sigma_1 = \begin{bmatrix} 1 & 0 \\ 0 & -1 \end{bmatrix}$$

$$\sigma_2 = \begin{bmatrix} 0 & 1 \\ 1 & 0 \end{bmatrix} \quad \sigma_3 = \begin{bmatrix} 0 & -i \\ i & 0 \end{bmatrix}$$

Mathematically, the factor of two relating physical angles to their counterparts in Stokes space derives from the use of second-order moments and correlations, and incorporates the loss of information due to absolute phase invariance.

The figure above makes use of a convenient representation of the last three Stokes parameters as components in a three-dimensional vector space. This space is closely related to the **Poincaré sphere**, which is the spherical surface occupied by completely polarized states in the space of the vector.

$$\mathbf{u} = \frac{1}{S_0}\begin{bmatrix} S_1 \\ S_2 \\ S_3 \end{bmatrix}$$

All four Stokes parameters can also be combined into the four-dimensional Stokes vector, which can be interpreted as four-vectors of Minkowski space. In this case, all physically realizable polarization states correspond to time-like, future-directed vectors.

Propagation, Reflection and Scattering

In a vacuum, the components of the electric field propagate at the speed of light, so that the phase of the wave varies in space in time while the polarization state does not. That is, $\mathbf{e}(z + \Delta z, t + \Delta t) = \mathbf{e}(z, t)e^{ik(c\Delta t - \Delta z)}$

where k is the wavenumber and positive z is the direction of propagation. As noted above, the physical electric vector is the real part of the Jones vector. When electromagnetic waves interact with matter, their propagation is altered. If this depends on the polarization states of the waves, then their polarization may also be altered.

In many types of media, electromagnetic waves may be decomposed into two orthogonal components that encounter different propagation effects. A similar situation occurs in the signal processing paths of detection systems that record the electric field directly. Such effects are most easily characterized in the form of a complex 2×2 transformation matrix called the Jones matrix:

é = JE

In general the Jones matrix of a medium depends on the frequency of the waves.

For propagation effects in two orthogonal modes, the Jones matrix can be written as

$$\mathbf{J} = \mathbf{T}\begin{bmatrix} g_1 & 0 \\ 0 & g_2 \end{bmatrix}\mathbf{T}^{-1}$$

where g_1 and g_2 are complex numbers representing the change in amplitude and phase caused in each of the two propagation modes, and **T** is a unitary matrix representing a change of basis from these propagation modes to the linear system used for the Jones vectors. For those media in which the amplitudes are unchanged but a differential phase delay occurs, the Jones matrix is unitary, while those affecting amplitude without phase have Hermitian Jones matrices. In fact, since any matrix may be written as the product of unitary and positive Hermitian matrices, any sequence of linear propagation effects, no matter how complex, can be written as the product of these two basic types of transformations.

Media in which the two modes accrue a differential delay are called birefringent. Well known manifestations of this effect appear in optical wave plates/retarders (linear modes) and in Faraday rotation/optical rotation (circular modes). An easily visualized example is one where the propagation modes are linear, and the incoming radiation is linearly polarized at a 45° angle to the modes. As the phase difference starts to appear, the polarization becomes elliptical, eventually changing to purely circular polarization (90° phase difference), then to elliptical and eventually linear polarization (180° phase) with an azimuth angle perpendicular to the original direction, then through circular again (270° phase), then elliptical with the original azimuth angle, and finally back to the original linearly polarized state (360° phase) where the cycle begins anew. In general the situation is more complicated and can be characterized as a rotation in the Poincaré sphere about the axis defined by the propagation modes (this is a consequence of the isomorphism of SU(2) with SO(3)). Examples for linear (blue), circular (red), and elliptical (yellow) birefringence are shown in the figure on the left. The total intensity and degree of polarization are unaffected. If the path length in the birefringent medium is sufficient, plane waves will exit the material with a significantly different propagation direction, due to refraction. For example, this is the case with macroscopic crystals of calcite, which present the viewer with two offset, orthogonally polarized images of whatever is viewed through them. It was this effect that provided the first discovery of polarization, by Erasmus Bartholinus in 1669. In addition, the phase shift, and thus the change in polarization state, is usually frequency dependent, which, in combination with dichroism, often gives rise to bright colors and rainbow-like effects.

Media in which the amplitude of waves propagating in one of the modes is reduced are called dichroic. Devices that block nearly all of the radiation in one mode are known as polarizing filters or simply "polarizers". In terms of the Stokes parameters, the total intensity is reduced while vectors in the Poincaré sphere are "dragged" towards the direction of the favored mode. Mathematically, under the treatment of the Stokes parameters as a Minkowski 4-vector, the transformation is a scaled Lorentz boost (due to the isomorphism of SL(2,C) and the restricted Lorentz group, SO(3,1)). Just as the Lorentz transformation preserves the proper time, the quantity det $\psi = S^2_n - S^2_1 - S^2_2 - S^2_3$ is invariant within a multiplicative scalar constant under Jones matrix transformations (dichroic and/or birefringent).

In birefringent and dichroic media, in addition to writing a Jones matrix for the net effect of passing through a particular path in a given medium, the evolution of the polarization state along that path can be characterized as the (matrix) product of an infinite series of infinitesimal steps, each operating on the state produced by all earlier matrices. In a uniform medium each step is the same, and one may write

$$\mathbf{J} = Je^{\mathbf{D}}$$

where J is an overall (real) gain/loss factor. Here **D** is a traceless matrix such that α**De** gives the derivative of e with respect to z. If **D** is Hermitian the effect is dichroism, while a unitary matrix models birefringence. The matrix **D** can be expressed as a linear combination of the Pauli matrices, where real coefficients give Hermitian matrices and imaginary coefficients give unitary matrices. The Jones matrix in each case may therefore be written with the convenient construction.

$$\mathbf{J_b} = J_b e^{\beta\sigma\text{-}\mathbf{n}} \text{ and } \mathbf{J_r} = J_r e^{\phi i \sigma\text{-}\mathbf{m}}$$

where σ is a 3-vector composed of the Pauli matrices (used here as generators for the Lie group SL(2, C)) and **n** and **m** are real 3-vectors on the Poincaré sphere corresponding to one of the propagation modes of the medium. The effects in that space correspond to a Lorentz boost of velocity parameter 2ß along the given direction, or a rotation of angle 2f about the given axis. These transformations may also be written as biquaternions (quaternions with complex elements), where the elements are related to the Jones matrix in the same way that the Stokes parameters are related to the coherency matrix. They may then be applied in pre- and post-multiplication to the quaternion representation of the coherency matrix, with the usual exploitation of the quaternion exponential for performing rotations and boosts taking a form equivalent to the matrix exponential equations above.

In addition to birefringence and dichroism in extended media, polarization effects describable using Jones matrices can also occur at (reflective) interface between two materials of different refractive index. These effects are treated by the Fresnel equations. Part of the wave is transmitted and part is reflected, with the ratio depending on angle of incidence and the angle of refraction. In addition, if the plane of the reflecting surface is not aligned with the plane of propagation of the wave, the polarization of the two parts is altered. In general, the Jones matrices of the reflection and transmission are real and diagonal, making the effect similar to that of a simple linear polarizer. For unpolarized light striking a surface at a certain optimum angle of incidence known as Brewster's angle, the reflected wave will be completely s-polarized.

Certain effects do not produce linear transformations of the Jones vector, and thus cannot be described with (constant) Jones matrices. For these cases it is usual instead to use a 4×4 matrix that acts upon the Stokes 4-vector. Such matrices were first used by Paul Soleillet in 1929, although they have come to be known as Mueller matrices. While every Jones matrix has a Mueller matrix, the reverse is not true. Mueller matrices are frequently used to study the effects of the scattering of waves from complex surfaces or ensembles of particles.

Polarization in Nature, Science, and Technology

Polarization Effects in Everyday Life

Light reflected by shiny transparent materials is partly or fully polarized, except when the light is normal (perpendicular) to the surface. It was through this effect that polarization was first discovered in 1808 by the mathematician Etienne Louis Malus. A polarizing filter, such as a pair of polarizing sunglasses, can be used to observe this effect by rotating the filter while looking through it at the reflection off of a distant horizontal surface. At certain rotation angles, the reflected light will be reduced or eliminated. Polarizing filters remove light polarized at 90° to the filter's polarization axis. If two polarizers are placed atop one another at 90° angles to one another, there is minimal light transmission.

Polarization by scattering is observed as light passes through the atmosphere. The scattered light produces the brightness and color in clear skies. This partial polarization of scattered light can be used to darken the sky in photographs, increasing the contrast. This effect is easiest to observe at sunset, on the horizon at a 90° angle from the setting sun. Another easily observed effect is the drastic reduction in brightness of images of the sky and clouds reflected from horizontal surfaces (see Brewster's angle), which is the main reason polarizing filters are often used in sunglasses. Also frequently visible through polarizing sunglasses are rainbow-like patterns caused by color-dependent birefringent effects, for example in toughened glass (e.g., car windows) or items made from transparent plastics. The role played by polarization in the operation of liquid crystal displays (LCDs) is also frequently apparent to the wearer of polarizing sunglasses, which may reduce the contrast or even make the display unreadable.

The photograph on the right was taken through polarizing sunglasses and through the rear window of a car.

Light from the sky is reflected by the windshield of the other car at an angle, making it mostly horizontally polarized. The rear window is made of tempered glass. Stress in the glass, left from its heat treatment, causes it to alter the polarization of light passing through it, like a wave plate. Without this effect, the sunglasses would block the horizontally polarized light reflected from the other car's window. The stress in the rear window, however, changes some of the horizontally polarized light into vertically polarized light that can pass through the glasses. As a result, the regular pattern of the heat treatment becomes visible.

CHAPTER 6

SOLAR ENERGY

Solar energy is the radiant light and heat from the Sun that has been harnessed by humans since ancient times using a range of ever-evolving technologies. Solar radiation along with secondary solar resources such as wind and wave power, hydroelectricity and biomass account for most of the available renewable energy on Earth. Only a minuscule fraction of the available solar energy is used.

Solar power technologies provide electrical generation by means of heat engines or photovoltaics. Once converted its uses are only limited by human ingenuity. A partial list of solar applications includes space heating and cooling through solar architecture, potable water via distillation and disinfection, daylighting, hot water, thermal energy for cooking, and high temperature process heat for industrial purposes.

Solar technologies are broadly characterized as either passive solar or active solar depending on the way they capture, convert and distribute sunlight. Active solar techniques include the use of photovoltaic panels, solar thermal collectors, with electrical or mechanical equipment, to convert sunlight into useful outputs. Passive solar techniques include orienting a building to the Sun, selecting materials with favorable thermal mass or light dispersing properties, and designing spaces that naturally circulate air.

Energy from the Sun

The Earth receives 174 petawatts (PW) of incoming solar radiation (insolation) at the upper atmosphere Approximately 30% is reflected back to space while the rest is absorbed by clouds, oceans and land masses. The spectrum of solar light at the Earth's surface is mostly spread across the visible and near-infrared ranges with a small part in the near-ultraviolet.

Earth's land surface, oceans and atmosphere absorb solar radiation, and this raises their temperature. Warm air containing evaporated water from the oceans rises, causing atmospheric circulation or convection. When the air reaches a high altitude, where the temperature is low, water vapour condenses into clouds, which rain onto the earth's surface, completing the water cycle. The latent heat of water condensation amplifies convection, producing atmospheric phenomena such as wind, cyclones and anti-cyclones. Sunlight absorbed by the oceans and land masses keeps the surface at an average temperature of 14 °C. By photosynthesis green plants convert solar energy into chemical energy, which produces food, wood and the biomass from which fossil fuels are derived.

The total solar energy absorbed by Earth's atmosphere, oceans and land masses is approximately 3,850,000 exajoules (EJ) per year. In 2002, this was more energy in one hour than the world used in one year Photosynthesis captures approximately 3,000 EJ per year in biomass The amount of solar energy reaching the surface of the planet is so vast that in one year it is about twice as much as will ever be obtained from all of the Earth's non-renewable resources of coal, oil, natural gas, and mined uranium combined.

From the table of resources it would appear that solar, wind or biomass would be sufficient to supply all of our energy needs, however, the increased use of biomass has had a negative effect on global warming and dramatically increased food prices by diverting forests and crops into biofuel production As intermittent resources, solar and wind raise other issues.

Applications of Solar Technology

Solar energy refers primarily to the use of solar radiation for practical ends. All other renewable energies other than geothermal derive their energy from the sun.

Solar technologies are broadly characterized as either passive or active depending on the way they capture, convert and distribute sunlight. Active solar techniques use photovoltaic panels, pumps, and fans to convert sunlight into useful outputs. Passive solar techniques include selecting materials with favorable thermal properties, designing spaces that naturally circulate air, and referencing the position of a building to the Sun. Active solar technologies increase the supply of energy and are considered supply side technologies, while passive solar technologies reduce the need for alternate resources and are generally considered demand side technologies.

Architecture and Urban Planning

Sunlight has influenced building design since the beginning of architectural historyAdvanced solar architecture and urban planning methods were first employed by the Greeks and Chinese, who oriented their buildings toward the south to provide light and warmth.

The common features of passive solar architecture are orientation relative to the Sun, compact proportion (a low surface area to volume ratio), selective shading (overhangs) and thermal mass. When these features are tailored to the local climate and environment they can produce well-lit spaces that stay in a comfortable temperature range. Socrates' Megaron House is a classic example of passive solar design The most recent approaches to solar design use computer modeling tying together solar lighting, heating and ventilation systems in an integrated solar design package Active solar equipment such as pumps, fans and switchable windows can complement passive design and improve system performance.

Urban heat islands (UHI) are metropolitan areas with higher temperatures than that of the surrounding environment. The higher temperatures are a result of increased absorption of the Solar light by urban materials such as asphalt and concrete, which have lower albedos and higher heat capacities than those in the natural environment. A straightforward method of counteracting the UHI effect is to paint buildings and roads white and plant trees. Using these methods, a hypothetical "cool communities" program in Los Angeles has projected that urban temperatures could be reduced by approximately 3 °C at an estimated cost of US$1 billion, giving estimated total annual benefits of US$530 million from reduced air-conditioning costs and healthcare savings.

Agriculture and Horticulture

Agriculture seeks to optimize the capture of solar energy in order to optimize the productivity of plants. Techniques such as timed planting cycles, tailored row orientation, staggered heights between rows and the mixing of plant varieties can improve crop yields While sunlight is generally considered a plentiful resource, the exceptions highlight the importance of solar energy to agriculture. During the short growing seasons of the Little Ice Age, French and English farmers employed fruit walls to maximize the collection of solar energy. These walls acted as thermal masses and accelerated ripening by keeping plants warm. Early fruit walls were built perpendicular to the ground and facing south, but over time, sloping walls were developed to make better use of sunlight. In 1699, Nicolas Fatio de Duillier even suggested using a tracking mechanism which could pivot to follow the Sun Applications of solar energy in agriculture aside from growing crops include pumping water, drying crops, brooding chicks and drying chicken manure More recently the technology has been embraced by vinters, who use the energy generated by solar panels to power grape presses.

Greenhouses convert solar light to heat, enabling year-round production and the growth (in enclosed environments) of specialty crops and other plants not naturally suited to the local climate. Primitive greenhouses were first used during Roman times to produce cucumbers year-round for the Roman emperor Tiberius The first modern greenhouses were built in Europe in the 16th century to keep exotic plants brought back from explorations abroad Greenhouses remain an important part of horticulture today, and plastic transparent materials have also been used to similar effect in polytunnels and row covers.

Solar Lighting

The history of lighting is dominated by the use of natural light. The Romans recognized a right to light as early as the 6th century and English law echoed these judgments with the Prescription Act of 1832 In the 20th century artificial lighting became the main source of interior illumination but daylighting techniques and hybrid solar lighting solutions are ways to reduce energy consumption.

Daylighting systems collect and distribute sunlight to provide interior illumination. This passive technology directly offsets energy use by replacing artificial lighting, and indirectly offsets non-solar energy use by reducing the need for air-conditioning Although difficult to quantify, the use of natural lighting also offers physiological and psychological benefits compared to artificial lighting. Daylighting design implies careful selection of window types, sizes and orientation; exterior shading devices may be considered as well. Individual features include sawtooth roofs, clerestory windows, light shelves, skylights and light tubes. They may be incorporated into existing structures, but are most effective when integrated into a solar design package that accounts for factors such as glare, heat flux and time-of-use. When daylighting features are properly implemented they can reduce lighting-related energy requirements by 25%.

Hybrid solar lighting is an active solar method of providing interior illumination. HSL systems collect sunlight using focusing mirrors that track the Sun and use optical fibers to transmit it inside the building to supplement conventional lighting. In single-story applications these systems are able to transmit 50% of the direct sunlight received.

Solar lights that charge during the day and light up at dusk are a common sight along walkways.

Although daylight saving time is promoted as a way to use sunlight to save energy, recent research has been limited and reports contradictory results: several studies report savings, but just as many suggest no effect or even a net loss, particularly when gasoline consumption is taken into account. Electricity use is greatly affected by geography, climate and economics, making it hard to generalize from single studies.

Solar Thermal

Solar thermal technologies can be used for water heating, space heating, space cooling and process heat generation.

Concentrating Solar Power

Solar troughs are the most widely deployed and the most cost-effective CSP technology.Concentrated sunlight has been used to perform useful tasks since the time of ancient China. A legend claims that Archimedes used polished shields to concentrate sunlight on the invading Roman fleet and repel them from Syracuse Auguste Mouchout used a parabolic trough to produce steam for the first solar steam engine in 1866, and subsequent developments led to the use of concentrating solar-powered devices for irrigation, refrigeration and locomotion.

Concentrating Solar Power (CSP) systems use lenses or mirrors and tracking systems to focus a large area of sunlight into a small beam. The concentrated light is then used as a heat source for a conventional power plant. A wide range of concentrating technologies exists; the most developed are the solar trough, parabolic dish and solar power tower. These methods vary in the way they track the Sun and focus light. In all these systems a working fluid is heated by the concentrated sunlight, and is then used for power generation or energy storage.

A solar trough consists of a linear parabolic reflector that concentrates light onto a receiver positioned along the reflector's focal line. The reflector is made to follow the Sun during the daylight hours by tracking along a single axis. Trough systems provide the best land-use factor of any solar technology The SEGS plants in California and Acciona's Nevada Solar One near Boulder City, Nevada are representatives of this technology.

A parabolic dish system consists of a stand-alone parabolic reflector that concentrates light onto a receiver positioned at the reflector's focal point. The reflector tracks the Sun along two axes. Parabolic dish systems give the highest efficiency among CSP technologies The 50 kW Big Dish in Canberra, Australia is an example of this technology The stirling solar dish combines a parabolic concentrating dish with a stirling heat engine which normally drives an electric generator. The advantages of stirling solar over photovoltaic cells are higher efficiency of converting sunlight into electricity and longer lifetime. A stirling engine has an approximate mean time before failure (MTBF) of 25 years.

A solar power tower uses an array of tracking reflectors (heliostats) to concentrate light on a central receiver atop a tower. Power towers are less advanced than trough systems but offer higher efficiency and better energy storage capability The Solar Two in Barstow, California and the Planta Solar 10 in Sanlucar la Mayor, Spain are representatives of this technology.

Experimental Solar Power

A solar updraft tower (also known as a solar chimney or solar tower) consists of a large greenhouse that funnels into a central tower. As sunlight shines on the greenhouse, the air inside is heated, and expands. The expanding air flows toward the central tower, where a turbine converts the air flow into electricity. A 50 kW prototype was constructed in Ciudad Real, Spain and operated for eight years before decommissioning in 1989.

A solar pond is a pool of salt water (usually 1-2 m deep) that collects and stores solar energy. Solar ponds were first proposed by Dr. Rudolph Bloch in 1948 after he came across reports of a lake in Hungary in which the temperature increased with depth. This effect was due to salts in the lake's water, which created a "density gradient" that prevented convection currents. A prototype was constructed in 1958 on the shores of the Dead Sea near Jerusalem The pond consisted of layers of water that successively increased from a weak salt solution at the top to a high salt solution at the bottom. This solar pond was capable of producing temperatures of 90° C in its bottom layer and had an estimated solar-to-electric efficiency of two percent.

Thermoelectric, or "thermovoltaic" devices convert a temperature difference between dissimilar materials into an electric current. First proposed as a method to store solar energy by solar pioneer Mouchout in the 1800s thermoelectrics reemerged in the Soviet Union during the 1930s. Under the direction of Soviet scientist Abram Ioffe a concentrating system was used to thermoelectrically generate power for a 1 hp engine Thermogenerators were later used in the US space program as an energy conversion technology for powering deep space missions such as Cassini, Galileo and Viking. Research in this area is focused on raising the efficiency of these devices from 7-8% to 15-20%.

Space solar power systems would use a large solar array in geosynchronous orbit to collect sunlight and beam this energy in the form of microwave radiation to receivers (rectennas) on Earth for distribution. This concept was first proposed by Dr. Peter Glaser in 1968 and since then a wide variety of systems have been studied with both photovoltaic and concentrating solar thermal technologies being proposed. Although still in the concept stage, these systems offer the possibility of delivering power approximately 96% of the time In 2008, John C. Mankins, a former NASA scientist, successfully used radio waves to send solar power between two Hawaiian islands in an experiment funded by the Discovery Channel. Mankins claims that this "proves the technology exists to beam solar power from satellites back to Earth."

Solar Chemical

Solar chemical processes use solar energy to drive chemical reactions. These processes offset energy that would otherwise come from an alternate source and can convert solar energy into storable and transportable fuels. Solar induced chemical reactions can be divided into thermochemical or photochemical.

Hydrogen production technologies been a significant area of solar chemical research since the 1970s. Aside from electrolysis driven by photovoltaic or photochemical cells, several thermochemical processes have also been explored. One such route uses concentrators to split water into oxygen and hydrogen at high temperatures (2300-2600 °C). Another approach uses the heat from solar concentrators to drive the steam reformation of natural gas thereby increasing the overall hydrogen yield compared to conventional reforming methods. Thermochemical cycles characterized by the decomposition and regeneration of reactants present another avenue for hydrogen production. The Solzinc process under development at the Weizmann Institute uses a 1 MW solar furnace to decompose zinc oxide (ZnO) at temperatures above 1200 °C. This initial reaction produces pure zinc, which can subsequently be reacted with water to produce hydrogen.

Sandia's Sunshine to Petrol (S2P) technology uses the high temperatures generated by concentrating sunlight along with a zirconia/ferrite catalyst to break down atmospheric carbon dioxide into oxygen and carbon monoxide (CO). The carbon monoxide can then be used to synthesize conventional fuels such as methanol, gasoline and jet fuel.

A photogalvanic device is a type of battery in which the cell solution (or equivalent) forms energy-rich chemical intermediates when illuminated. These energy-rich intermediates can potentially be stored and subsequently reacted at the electrodes to produce an electric potential. The ferric-thionine chemical cell is an example of this technology.

Photoelectrochemical cells or PECs consist of a semiconductor, typically titanium dioxide or related titanates, immersed in an electrolyte. When the semiconductor is illuminated an electrical potential develops. There are two types of photoelectrochemical cells: photoelectric cells that convert light into electricity and photochemical cells that use light to drive chemical reactions such as electrolysis.

Energy Storage Methods

Solar energy is not available at night, and energy storage is an important issue because modern energy systems usually assume continuous availability of energy.

Thermal mass systems can store solar energy in the form of heat at domestically useful temperatures for daily or seasonal durations. Thermal storage systems generally use readily available materials with high specific heat capacities such as water, earth and stone. Well-designed systems can lower peak demand, shift time-of-use to off-peak hours and reduce overall heating and cooling requirements.

Phase change materials such as paraffin wax and Glauber's salt are another thermal storage media. These materials are inexpensive, readily available, and can deliver domestically useful temperatures (approximately 64 °C). The "Dover House" (in Dover, Massachusetts) was the first to use a Glauber's salt heating system, in 1948.

Solar energy can be stored at high temperatures using molten salts. Salts are an effective storage medium because they are low-cost, have a high specific heat capacity and can deliver heat at temperatures compatible with conventional power systems. The Solar Two used this method of energy storage, allowing it to store 1.44 TJ in its 68 m^3 storage tank with an annual storage efficiency of about 99%.

Off-grid PV systems have traditionally used rechargeable batteries to store excess electricity. With grid-tied systems, excess electricity can be sent to the transmission grid. Net metering programs give these systems a credit for the electricity they deliver to the grid. This credit offsets electricity provided from the grid when the system cannot meet demand, effectively using the grid as a storage mechanism.

Pumped-storage hydroelectricity stores energy in the form of water pumped when energy is available from a lower elevation reservoir to a higher elevation one. The energy is recovered when demand is high by releasing the water to run through a hydroelectric power generator.

Development, Deployment and Economics

Beginning with the surge in coal use which accompanied the Industrial Revolution, energy consumption has steadily transitioned from wood and biomass to fossil fuels. The early development of solar technologies starting in the 1860s was driven by an expectation that coal would soon become scarce. However development of solar technologies stagnated in the early 20th century in the face of the increasing availability, economy, and utility of coal and petroleum.

The 1973 oil embargo and 1979 energy crisis caused a reorganization of energy policies around the world and brought renewed attention to developing solar technologies Deployment strategies focused on incentive programs such as the Federal Photovoltaic Utilization Program in the US and the Sunshine Program in Japan. Other efforts included the formation of research facilities in the US (SERI, now NREL), Japan (NEDO), and Germany (Fraunhofer Institute for Solar Energy Systems ISE

Between 1970 and 1983 photovoltaic installations grew rapidly, but falling oil prices in the early 1980s moderated the growth of PV from 1984 to 1996. Since 1997, PV development has accelerated due to supply issues with oil and natural gas, global warming concerns (see Kyoto Protocol), and the improving economic position of PV relative to other energy technologies.[citation needed] Photovoltaic production growth has averaged 40% per year since 2000 and installed capacity reached 10.6 GW at the end of 2007. Since 2006 it has been economical for investors to install photovoltaics for free in return for a long term power purchase agreement. 50% of commercial systems were installed in this manner in 2007 and it is expected that 90% will by 2009 Nellis Air Force Base is receiving photoelectric power for about 2.2 ¢/kWh and grid power for 9 ¢/kWh.

Commercial solar water heaters began appearing in the United States in the 1890s These systems saw increasing use until the 1920s but were gradually replaced by cheaper and more reliable heating fuels As with photovoltaics, solar water heating attracted renewed attention as a result of the oil crises in the 1970s but interest subsided in the 1980s due to falling petroleum prices. Development in the solar water heating sector progressed steadily throughout the 1990s and growth rates have averaged 20% per year since 1999 Although generally underestimated, solar water heating is by far the most widely deployed solar technology with an estimated capacity of 154 GW as of 2007.

Commercial concentrating solar power (CSP) plants were first developed in the 1980s. CSP plants such as SEGS project in the United States have a LEC of 12–14 ¢/kWh. The 11 MW PS10 power tower in Spain, completed in late 2005, is Europe's first commercial CSP system, and a total capacity of 300 MW is expected to be installed in the same area by 2013.

Solar installations in recent years have also largely begun to expand into residential areas, with governments offering incentive programs to make "green" energy a more economically viable option. In Canada the government offers the RESOP (Renewable Energy Standard Offer Program). The program allows residential homeowners with solar panel installations to sell the energy they produce back to the grid (i.e., the government) at 41¢/kWh, while drawing power from the grid at an average rate of 20¢/kWh . The program is designed to help promote the government's green agenda and lower the strain often placed on the energy grid at peak hours. With the incentives offered by the program the average payback period for a residential solar installation (sized between 1.3 kW and 5 kW) is estimated at 18 to 23 years, considering such cost factors as parts, installation and maintenance, as well as the average energy production of a system on an annual basis.

Daniel Lincot, the chairman of the 2008 European Photovoltaic Solar Energy Conference and the research director of the Paris-based Photovoltaic Energy Development and Research Institute, said that photovoltaics can cover all the world energy demand [dead link]. Photovoltaics are 85 times as efficient as growing corn for ethanol. On a 300 feet (91 m) by 300 feet (1 hectare) plot of land enough ethanol can be produced to drive a car 30,000 miles (48,000 km) per year or 2,500,000 miles (4,020,000 km) by covering the same land with photo cells. The deserts of the South Western United States could produce sufficient electricity to fulfill all of the electrical needs of the United States, and could even electrolyze water into Hydrogen and Oxygen to power the entire U.S. land fleet.

SUNLIGHT

Sunlight, in the broad sense, is the total spectrum of the electromagnetic radiation given off by the Sun. On Earth, sunlight is filtered through the atmosphere, and the solar radiation is obvious as daylight when the Sun is above the horizon. Near the poles in summer, the days are longer and the nights are shorter t from the increase in atmospheric temperature due to the radiative heating of the atmosphere by the sun's radiation. Sunlight may be recorded using a sunshine recorder, pyranometer or pyr shine as direct irradiance from the Sun measured on the ground of at least 120 $W \cdot m^{-2}$ light. Bright sunlight provides luminance of approximately 100,000 candela per square meter at the Earth's surface. Sunlight is a key factor in the process of photosynthesis.

Calculation

To calculate the amount of sunlight reaching the ground, both the ellipt lluminance (Eext), corrected for the elliptical orbit by using the day number of the year (dn), is:

$$E_{ext} = \frac{E_{sc}}{\left[1+0.034.\cos\left(2\pi\frac{dn-3}{365}\right)\right]}$$

where dn = 1 on January 1; dn = 2 January 2; dn = 32 on February 1, etc. In this formula dn-3 is used, because in modern times Earth's perihelion, the closest approach to the Sun and therefore the maximum Eext, occurs around January 3 each year.

The solar illuminance constant (E_{sc}), is equal to 128×103 lx. The direct normal illuminance (E_{dn}), corrected for the attenuating effects of the atmosphere is given by:

$$E_{dn} = E_{ext}\, e^{-cm},$$

where c is the atmospheric extinction coefficient and m is the relative optical airmass.

Solar Constant

The solar constant is the amount of incoming solar electromagnetic radiation per unit area, measured on the outer surface of Earh's atmosphere in a plane perpendicular to the rays. The solar constant includes all types of solar radiation, not just the visible light. It is measured by satellite to be roughly 1,366 watts per square meter (W/m^2) though this fluctuates by about 6.9% during a year (from 1,412 W/m^2 in early January to 1,321 W/m^2 in early July) due to the earth's varying distance from the Sun, and typically by much less than one part per thousand from day to day. Thus, for the whole Earth (which has a cross section of 127,400,000 km^2), the power is 1.740 × 1017 W, plus or minus 3.5%. The solar constant does not remain constant over long periods of time (see Solar variation). The approximate average value cited 1,366 W/m^2, is equivalent to 1.96 calories per minute per square centimeter, or 1.96 langleys (Ly) per minute.

The Earth receives a total amount of radiation determined by its cross section (p·RE^2), but as it rotates this energy is distributed across the entire surface area (4·p·RE^2). Hence the average incoming solar radiation (sometimes called the solar irradiance), taking into account the angle at which the rays strike and that at any one moment half the planet does not receive any solar radiation, is one-fourth the solar constant (approximately 342 W/m^2). At any given moment, the amount of Solar radiation received at a location on the Earth's surface depends on the state of the atmosphere and the location's latitude.

The solar constant includes all wavelengths of solar electromagnetic radiation, not just the visible light (see Electromagnetic spectrum). It is linked to the apparent magnitude of the Sun, -26.8, in that the solar constant and the magnitude of the Sun are two methods of describing the apparent brightness of the Sun, though the magnitude only measures the visual output of the Sun.

In 1884, Samuel Pierpont Langley attempted to estimate the Solar constant from Mount Whitney in California. By taking readings at different times of day, he attempted to remove effects due to atmospheric absorption. However, the value he obtained, 2,903 W/m^2, was still too great. Between 1902 and 1957, measurements by Charles Greeley Abbot and others at various high-altitude sites found values between 1,322 and 1,465 W/m^2. Abbott proved that one of Langley's corrections was erroneously applied. His results varied between 1.89 and 2.22 calories (1,318 to 1,548 W/m^2), a variation that appeared to be due to the Sun and not the Earth's atmosphere.

The angular diameter of the Earth as seen from the Sun is approximately 1/11,000 radians, meaning the solid angle of the Earth as seen from the sun is approximately 1/140,000,000 steradians. Thus the Sun emits about two billion times the amount of radiation that is caught by Earth, in other words about 3.86×1026 watts.

Sunlight Intensity in the Solar System

The actual brightness of sunlight that would be observed at the surface depends also on the presence and composition of an atmosphere. For example Venus' thick atmosphere reflects more than 60% of the solar light it receives. The actual illumination of the surface is about 5,000-10,000 lux, comparable to that of Earth during a dark, very cloudy day.

Sunlight on Mars would be more or less like daylight on Earth wearing sunglasses, and as can be seen in the pictures taken by the rovers, there is enough diffuse sky radiation that shadows would not seem particularly dark. Thus it would give perceptions and "feel" very much like Earth daylight.

For comparison purposes, sunlight on Saturn is somewhat slightly brighter than Earth sunlight on the average sunset or sunrise. Even on Pluto the Sun would be still bright enough to almost match the average living room. To see the Sun shine as dim as the full Moon on the Earth, a distance of about 500 AU (~69 light-hours) is needed: there is only a handful of objects in the solar system known to orbit farther than such a distance, among them 90377 Sedna and (87269) 2000 OO67.

Composition

The spectrum of the Sun's solar radiation is close to that of a black body with a temperature of about 5,800 K. About half that lies in the visible short-wave part of the electromagnetic spectrum and the other half mostly in the near-infrared part. Some also lies in the ultraviolet part of the spectrum When ultraviolet radiation is not absorbed by the atmosphere or other protective coating, it can cause a change in human skin pigmentation.

The spectrum of electromagnetic radiation striking the Earth's atmosphere is 100 to 106 nanometers (nm). This can be divided into five regions in increasing order of wavelengths.

- *Ultraviolet C* or (UVC) range, which spans a range of 100 to 280 nm. The term ultraviolet refers to the fact that the radiation is at higher frequency than violet light (and, hence also invisible to the human eye). Owing to absorption by the atmosphere very little reaches the Earth's surface (Lithosphere). This spectrum of radiation has germicidal properties, and is used in germicidal lamps.
- Ultraviolet *B* or (UVB) range spans 280 to 315 nm. It is also greatly absorbed by the atmosphere, and along with UVC is responsible for the photochemical reaction leading to the production of the Ozone layer.
- *Ultraviolet A* or (UVA) spans 315 to 400 nm. It has been traditionally held as less damaging to the DNA, and hence used in tanning and PUVA therapy for psoriasis.
- Visible range or light spans 400 to 700 nm. As the name suggests, it is this range that is visible to the naked eye.
- *Infrared* range that spans 700 nm to 106 nm [1 (mm)]. It is largely responsible for the warmth or heat that the sunlight carries. It is also divided into three types on the basis of wavelength:
  - Infrared-A: 700 nm to 1,400 nm
  - Infrared-B: 1,400 nm to 3,000 nm
  - Infrared-C: 3,000 nm to 1 mm.

Climate Effects

On Earth, solar radiation is obvious as daylight when the sun is above the horizon. This is during daytime, and also in summer near the poles at night, but not at all in winter near the poles. When the direct radiation is not blocked by clouds, it is experienced as sunshine, combining the perception of bright white light (sunlight in the strict sense) and warming. The warming on the body and surfaces of other objects is distinguished from the increase in air temperature.

The amount of radiation intercepted by a planetary body varies inversely with the square of the distance between the star and the planet. The Earth's orbit and obliquity change with time (over thousands of years), sometimes forming a nearly perfect circle, and at other times stretching out to an orbital eccentricity of 5% (currently 1.67%). The total insolation remains almost constant but the seasonal and latitudinal distribution and intensity of solar radiation received at the Earth's surface also varies For example, at latitudes of 65 degrees the change in solar energy in summer & winter can vary by more than 25% as a result of the Earth's orbital variation. Because changes in winter and summer tend to offset, the change in the annual average insolation at any given location is near zero, but the redistribution of energy between summer and winter does strongly affect the intensity of seasonal cycles. Such changes associated with the redistribution of solar energy are considered a likely cause for the coming and going of recent ice ages.

Life on Earth

The existence of nearly all life on Earth is fueled by light from the sun. Most autotrophs, such as plants, use the energy of sunlight to turn air into simple sugars—a process known as photosynthesis. These sugars are then used as building blocks and in other synthetic pathways which allow the organism to grow.

Heterotrophs, such as animals, use light from the sun indirectly by consuming the products of autotrophs, either directly or by consuming other heterotrophs. The sugars and other molecular components produced by the autotrophs are then broken down, releasing stored solar energy, and giving the heterotroph the energy required for survival. This process is known as respiration.

In prehistory, humans began to further extend this process by putting plant and animal materials to other uses. They used animal skins for warmth, for example, or wooden weapons to hunt. These skills allowed humans to harvest more of the sunlight than was possible through glycolysis alone, and human population began to grow.

During the Neolithic Revolution, the domestication of plants and animals further increased human access to solar energy. Fields devoted to crops were enriched by inedible plant matter, providing sugars and nutrients for future harvests. Animals which had previously only provided humans with meat and tools once they were killed were now used for labour throughout their lives, fueled by grasses inedible to humans.

The more recent discoveries of coal, petroleum and natural gas are modern extensions of this trend. These fossil fuels are the remnants of ancient plant and animal matter, formed using energy from sunlight and then trapped within the earth for millions of years. Because the stored energy in these fossil fuels has accumulated over many millions of years, they have allowed modern humans to massively increase the production and consumption of primary energy. As the amount of fossil fuel is large but finite, this cannot continue indefinitely, and various theories exist as to what will follow this stage of human civilization (e.g. alternative fuels, Malthusian catastrophe, new urbanism, peak oil).

INFRARED

Infrared (IR) radiation is electromagnetic radiation whose wavelength is longer than that of visible light (400-700 nm), but shorter than that of terahertz radiation (3-300 μm) and microwaves (~30,000 um). Infrared radiation spans roughly three orders of magnitude (750 nm and 1000 μm).

Direct sunlight has a luminous efficacy of about 93 lumens per watt of radiant flux, which includes infrared (47% share of the spectrum), visible (46%), and ultra-violet (only 7%) light. Bright sunlight provides luminance of approximately 100,000 candela per square meter at the Earth's surface.

Overview

Infrared imaging is used extensively for both military and civilian purposes. Military applications include target acquisition, surveillance, night vision, homing and tracking. Non-military uses include thermal efficiency analysis, remote temperature sensing, short-ranged wireless communication, spectroscopy, and weather forecasting. Infrared astronomy uses sensor-equipped telescopes to penetrate dusty regions of space, such as molecular clouds; detect cool objects such as planets, and to view highly red-shifted objects from the early days of the universe. Humans at normal body temperature radiate chiefly at wavelengths around 10μm (micrometers).

At the atomic level, infrared energy elicits vibrational modes in a molecule through a change in the dipole moment, making it a useful frequency range for study of these energy states. Infrared spectroscopy examines absorption and transmission of photons in the infrared energy range, based on their frequency and intensity.

Origins of the Term

The name means below red (from the Latin infra, "below"), red being the color of the longest wavelengths of visible light. IR light has a longer wavelength (a lower frequency) than that of red light, hence below.

Different Regions in the Infrared

Objects generally emit infrared radiation across a spectrum of wavelengths, but only a specific region of the spectrum is of interest because sensors are usually designed only to collect radiation within a specific bandwidth. As a result, the infrared band is often subdivided into smaller sections.

CIE Division Scheme

The International Commission on Illumination (CIE) recommended the division of optical radiation into the following three bands:

1. IR-A: 700 nm–1400 μm
2. IR-B: 1400 nm–3000 μm
3. IR-C: 3000 nm–1 μm

A commonly used sub-division scheme is:

- *Near-infrared (NIR, IR-A DIN):* 0.75-1.4 μm in wavelength, defined by the water absorption, and commonly used in fiber optic telecommunication because of low attenuation losses in the SiO2 glass (silica) medium. Image intensifiers are sensitive to this area of the spectrum. Examples include night vision devices such as night vision goggles.
- *Short-wavelength infrared(SWIR, IR-B DIN):* 1.4-3 μm, water absorption increases significantly at 1,450 nm. The 1,530 to 1,560 nm range is the dominant spectral region for long-distance telecommunications.
- *Mid-wavelength infrared (MWIR, IR-C DIN) also called intermediate infrared (IIR):* 3-8 μm. In guided missile technology the 3-5 μm portion of this band is the atmospheric window in which the homing heads of passive IR 'heat seeking' missiles are designed to work, homing on to the IR signature of the target aircraft, typically the jet engine exhaust plume.
- *Long-wavelength infrared (LWIR, IR-C DIN):* 8–15 μm. This is the "thermal imaging" region, in which sensors can obtain a completely passive picture of the outside world based on thermal emissions only and requiring no external light or thermal source such as the sun, moon or infrared illuminator. Forward-looking infrared (FLIR) systems use this area of the spectrum. Sometimes also called the "far infrared."
- *Far infrared (FIR): 15-1,000 μm:*NIR and SWIR is sometimes called reflected infrared while MWIR and LWIR is sometimes referred to as thermal infrared. Due

to the nature of the blackbody radiation curves, typical 'hot' objects, such as exhaust pipes, often appear brighter in the MW compared to the same object viewed in the LW.

Astronomers typically divide the infrared spectrum as follows:

- near: (0.7-1) to 5 μm
- mid: 5 to (25-40) μm
- long: (25-40) to (200-350) μm

These divisions are not precise and can vary depending on the publication. The three regions are used for observation of different temperature ranges, and hence different environments in space.

A third scheme divides up the band based on the response of various detectors.

- *Near infrared (NIR):* from 0.7 to 1.0 micrometers (from the approximate end of the response of the human eye to that of silicon).
- *Short-wave infrared (SWIR):* 1.0 to 3 micrometers (from the cut off of silicon to that of the MWIR atmospheric window. InGaAs covers to about 1.8 micrometers; the less sensitive lead salts cover this region.
- *Mid-wave infrared (MWIR):* 3 to 5 micrometers (defined by the atmospheric window and covered by Indium antimonide [InSb] and HgCdTe and partially by lead selenide [PbSe]).
- *Long-wave infrared (LWIR):* 8 to 12, or 7 to 14 micrometers: the atmospheric window (Covered by HgCdTe and microbolometers).
- *Very-long wave infrared (VLWIR):* 12 to about 30 micrometers, covered by doped silicon.

These divisions are justified by the different human response to this radiation: near infrared is the region closest in wavelength to the radiation detectable by the human eye, mid and far infrared are progressively further from the visible regime. Other definitions follow different physical mechanisms (emission peaks, vs. bands, water absorption) and the newest follow technical reasons (The common silicon detectors are sensitive to about 1,050 nm, while InGaAs' sensitivity starts around 950 nm and ends between 1,700 and 2,600 nm, depending on the specific configuration). Unfortunately, international standards for these specifications are not currently available.

The boundary between visible and infrared light is not precisely defined. The human eye is markedly less sensitive to light above 700 nm wavelength, so shorter frequencies make insignificant contributions to scenes illuminated by common light sources. But particularly intense light (e.g., from lasers, or from bright daylight with the visible light removed by colored gels can be detected up to approximately 780 nm, and will be perceived as red light. The onset of infrared is defined (according to different standards) at various values typically between 700 nm and 800 nm.

Telecommunication Bands in the Infrared

In optical communications, the part of the infrared spectrum that is used is divided into several bands based on availability of light sources, transmitting/absorbing materials (fibers) and detectors.

The C-band is the dominant band for long-distance telecommunication networks. The S and L bands are based on less well established technology, and are not as widely deployed.

Heat

Infrared radiation is popularly known as "heat" or sometimes "heat radiation", since many people attribute all radiant heating to infrared light and/or to all infrared radiation to being a result of heating. This is a widespread misconception, since light and electromagnetic waves of any frequency will heat surfaces that absorb them. Infrared light from the Sun only accounts for 49% of the heating of the Earth, with the rest being caused by visible light that is absorbed then re-radiated at longer wavelengths. Visible light or ultraviolet-emitting lasers can char paper and incandescently hot objects emit visible radiation. It is true that objects at room temperature will emit radiation mostly concentrated in the 8 to 12 micrometer band, but this is not distinct from the emission of visible light by incandescent objects and ultraviolet by even hotter objects.

Heat is energy in transient form that flows due to temperature difference. Unlike heat transmitted by thermal conduction or thermal convection, radiation can propagate through a vacuum.

The concept of emissivity is important in understanding the infrared emissions of objects. This is a property of a surface which describes how its thermal emissions deviate from the ideal of a black body. To further explain, two objects at the same physical temperature will not 'appear' the same temperature in an infrared image if they have differing emissivities.

Infrared (IR) filters can be made from many different materials. One type is made of polysulphone plastic that blocks over 99% of the visible light spectrum from "white" light sources such as incandescent filament bulbs. Infrared filters allow a maximum of infrared output while maintaining extreme covertness. Currently in use around the world, infrared filters are used in Military, Law Enforcement, Industrial and Commercial applications. The unique makeup of the plastic allows for maximum durability and heat resistance. IR filters provide a more cost effective and time efficient solution over the standard bulb replacement alternative. All generations of night vision devices are greatly enhanced with the use of IR filters.

Applications

Night Vision

Infrared is used in night vision equipment when there is insufficient visible light to see. Night vision devices operate through a process involving the conversion of ambient light photons into electrons which are then amplified by a chemical and electrical process and then converted back into visible light. Infrared light sources can be used to augment the available ambient light for conversion by night vision devices, increasing in-the-dark visibility without actually using a visible light source.

The use of infrared light and night vision devices should not be confused with thermal imaging which creates images based on differences in surface temperature by detecting infrared radiation (heat) that emanates from objects and their surrounding environment.

Thermography

Infrared radiation can be used to remotely determine the temperature of objects (if the emissivity is known). This is termed thermography, or in the case of very hot objects in the NIR or visible it is termed pyrometry. Thermography (thermal imaging) is mainly used in military and industrial applications but the technology is reaching the public market in the form of infrared cameras on cars due to the massively reduced production costs.

Thermographic cameras detect radiation in the infrared range of the electromagnetic spectrum (roughly 900–14,000 nanometers or 0.9-14 μm) and produce images of that radiation. Since infrared radiation is emitted by all objects based on their temperatures, according to the black body radiation law, thermography makes it possible to "see" one's environment with or without visible illumination. The amount of radiation emitted by an object increases with temperature, therefore thermography allows one to see variations in temperature (hence the name).

Other Imaging

In *infrared photography, infrared filters* are used to capture the near-infrared spectrum. Digital cameras often use infrared blockers. Cheaper digital cameras and camera phones have less effective filters and can "see" intense near-infrared, appearing as a bright purple-white color. This is especially pronounced when taking pictures of subjects near IR-bright areas (such as near a lamp), where the resulting infrared interference can wash out the image. There is also a technique called 'T-ray' imaging, which is imaging using far infrared or terahertz radiation. Lack of bright sources makes terahertz photography technically more challenging than most other infrared imaging techniques. Recently T-ray imaging has been of considerable interest due to a number of new developments such as terahertz time-domain spectroscopy.

Tracking

Infrared tracking, also known as infrared homing, refers to a passive missile guidance system which uses the emission from a target of electromagnetic radiation in the infrared part of the spectrum to track it. Missiles which use infrared seeking are often referred to as "heat-seekers", since infrared (IR) is just below the visible spectrum of light in frequency and is radiated strongly by hot bodies. Many objects such as people, vehicle engines and aircraft generate and retain heat, and as such, are especially visible in the infra-red wavelengths of light compared to objects in the background.

Heating

Infrared radiation can be used as a deliberate heating source. For example it is used in infrared saunas to heat the occupants, and also to remove ice from the wings of aircraft (de-icing). FIR is also gaining popularity as a safe method of natural health care & physiotherapy. Far infrared thermomedic therapy garments use thermal technology to provide compressive support and healing warmth to assist symptom control for arthritis, injury & pain. Infrared can be used in cooking and heating food as it predominantly heats the opaque, absorbent objects, rather than the air around them.

Infrared heating is also becoming more popular in industrial manufacturing processes, e.g. curing of coatings, forming of plastics, annealing, plastic welding, print drying. In these applications, infrared heaters replace convection ovens and contact heating. Efficiency is achieved by matching the wavelength of the infrared heater to the absorption characteristics of the material.

Communications

IR data transmission is also employed in short-range communication among computer peripherals and personal digital assistants. These devices usually conform to standards published by IrDA, the Infrared Data Association. Remote controls and IrDA devices use infrared light-emitting diodes (LEDs) to emit infrared radiation which is focused by a plastic lens into a narrow beam. The beam is modulated, i.e. switched on and off, to encode the data. The receiver uses a silicon photodiode to convert the infrared radiation to an electric current. It responds only to the rapidly pulsing signal created by the transmitter, and filters out slowly changing infrared radiation from ambient light. Infrared communications are useful for indoor use in areas of high population density. IR does not penetrate walls and so does not interfere with other devices in adjoining rooms. Infrared is the most common way for remote controls to command appliances. Free space optical communication using infrared lasers can be a relatively inexpensive way to install a communications link in an urban area operating at up to 4 gigabit/s, compared to the cost of burying fiber optic cable. Infrared lasers are used to provide the light for optical fiber communications systems. Infrared light with a wavelength around 1,330 nm (least dispersion) or 1,550 nm (best transmission) are the best choices for standard silica fibers. IR data transmission of encoded audio versions of printed signs is being researched as an aid for visually impaired people through the RIAS (Remote Infrared Audible Signage) project.

Spectroscopy

Infrared vibrational spectroscopy (see also near infrared spectroscopy) is a technique which can be used to identify molecules by analysis of their constituent bonds. Each chemical bond in a molecule vibrates at a frequency which is characteristic of that bond. A group of atoms in a molecule (e.g. CH_2) may have multiple modes of oscillation caused by the stretching and bending motions of the group as a whole. If an oscillation leads to a change in dipole in the molecule, then it will absorb a photon which has the same frequency. The vibrational frequencies of most molecules correspond to the frequencies of infrared light. Typically, the technique is used to study organic compounds using light radiation from 4000-400 cm^{-1}, the mid-infrared. A spectrum of all the frequencies of absorption in a sample is recorded. This can be used to gain information about the sample composition in terms of chemical groups present and also its purity (for example a wet sample will show a broad O-H absorption around 3200 cm^{-1}).

Meteorology

Weather satellites equipped with scanning radiometers produce thermal or infrared images which can then enable a trained analyst to determine cloud heights and types, to calculate land and surface water temperatures, and to locate ocean surface features. The scanning is typically in the range 10.3-12.5 μm (IR4 and IR5 channels).

High, cold ice cloud such as *Cirrus* or *Cumulonimbus* show up bright white, lower warmer cloud such as Stratus or *Stratocumulus* show up as grey with intermediate clouds shaded accordingly. Hot land surfaces will show up as dark grey or black. One disadvantage of infrared imagery is that low cloud such as stratus or fog can be a similar temperature to the surrounding land or sea surface and does not show up. However, using the difference in brightness of the IR4 channel (10.3-11.5 μm) and the near-infrared channel (1.58-1.64 μm), low cloud can be distinguished, producing a *fog* satellite picture. The main advantage of infrared is that images can be produced at night, allowing a continuous sequence of weather to be studied.

These infrared pictures can depict ocean eddies or vortices and map currents such as the Gulf Stream which are valuable to the shipping industry. Fishermen and farmers are interested in knowing land and water temperatures to protect their crops against frost or increase their catch from the sea. Even El Niño phenomena can be spotted. Using color-digitized techniques, the gray shaded thermal images can be converted to color for easier identification of desired information.

Climatology

In the field of climatology, atmospheric infrared radiation is monitored to detect trends in the energy exchange between the earth and the atmosphere. These trends provide information on long term changes in the earth's climate. It is one of the primary parameters studied in research into global warming together with solar radiation.

A pyrgeometer is utilized in this field of research to perform continuous outdoor measurements. This is a broadband infrared radiometer with sensitivity for infrared radiation between approximately 4.5 μm and 50 μm.

Astronomy

Astronomers observe objects in the infrared portion of the electromagnetic spectrum using optical components, including mirrors, lenses and solid state digital detectors. For this reason it is classified as part of optical astronomy. To form an image, the components of an infrared telescope need to be carefully shielded from heat sources, and the detectors are chilled using liquid helium.

The sensitivity of Earth-based infrared telescopes is significantly limited by water vapor in the atmosphere, which absorbs a portion of the infrared radiation arriving from space outside of selected atmospheric windows. This limitation can be partially alleviated by placing the telescope observatory at a high altitude, or by carrying the telescope aloft with a balloon or an aircraft. Space telescopes do not suffer from this handicap, and so outer space is considered the ideal location for infrared astronomy.

The infrared portion of the spectrum has several useful benefits for astronomers. Cold, dark molecular clouds of gas and dust in our galaxy will glow with radiated heat as they are irradiated by imbedded stars. Infrared can also be used to detect protostars before they begin to emit visible light. Stars emit a smaller portion of their energy in the infrared spectrum, so nearby cool objects such as planets can be more readily detected. (In the visible light spectrum, the glare from the star will drown out the reflected light from a planet.)

Infrared light is also useful for observing the cores of active galaxies which are often cloaked in gas and dust. Distant galaxies with a high redshift will have the peak portion of their spectrum shifted toward longer wavelengths, so they are more readily observed in the infrared.

CHAPTER 7 ELECTROMAGNETIC RADIATION

Electromagnetic (EM) radiation (sometimes abbreviated EMR) takes the form of self-propagating waves in a vacuum or in matter. EM radiation has an electric and magnetic field component which oscillate in phase perpendicular to each other and to the direction of energy propagation. Electromagnetic radiation is classified into types according to the frequency of the wave, these types include (in order of increasing frequency): radio waves, microwaves, terahertz radiation, infrared radiation, visible light, ultraviolet radiation, X-rays and gamma rays. Of these, radio waves have the longest wavelengths and Gamma rays have the shortest. A small window of frequencies, called visible spectrum or light, is sensed by the eye of various organisms, with variations of the limits of this narrow spectrum.

The EM radiation carries energy and momentum, which may be imparted when it interacts with matter.

Physics

Theory

Electromagnetic waves were first postulated by James Clerk Maxwell and subsequently confirmed by Heinrich

Hertz. Maxwell derived a wave form of the electric and magnetic equations, revealing the wave-like nature of electric and magnetic fields, and their symmetry. Because the speed of EM waves predicted by the wave equation coincided with the measured speed of light, Maxwell concluded that light itself is an EM wave.

According to *Maxwell's equations*, a time-varying electric field generates a magnetic field and vice versa. Therefore, as an oscillating electric field generates an oscillating magnetic field, the magnetic field in turn generates an oscillating electric field, and so on. These oscillating fields together form an electromagnetic wave.

A *quantum theory* of the interaction between electromagnetic radiation and matter such as electrons is described by the theory of quantum electrodynamics.

Properties

Electric and magnetic fields do obey the properties of superposition, so fields due to particular particles or time-varying electric or magnetic fields contribute to the fields due to other causes. (As these fields are vector fields, all magnetic and electric field vectors add together according to vector addition.) These properties cause various phenomena including refraction and diffraction. For instance, a travelling EM wave incident on an atomic structure induces oscillation in the atoms, thereby causing them to emit their own EM waves. These emissions then alter the impinging wave through interference.

Since light is an oscillation, it is not affected by travelling through static electric or magnetic fields in a linear medium such as a vacuum. In nonlinear media such as some crystals, however, interactions can occur between light and static electric and magnetic fields-these interactions include the Faraday effect and the Kerr effect.

In refraction, a wave crossing from one medium to another of different density alters its speed and direction upon entering the new medium. The ratio of the refractive indices of the media determines the degree of refraction, and is summarized by Snell's law. Light disperses into a visible spectrum as light is shone through a prism because of the wavelength dependant refractive index of the prism material (Dispersion).

The physics of electromagnetic radiation is electrodynamics, a subfield of electromagnetism. The EM radiation exhibits both wave properties and particle properties at the same time (see wave-particle duality). The wave charactnces and timescales. Both characteristics have been confirmed in a large number of experiments.

There are experiments in which the wave and particle natures of electromagnetic waves appear in the same experiment, such as the diffraction of a single photon. When a single photon is sent through two slits, it passes through both of them interfering with itself, as waves do, yet is detected by a photomultiplier or other sesitive detector only once. Similar self-interference is observed when a single photon is sent into a Michelson interferometer or other interferometers.

Wave Model

An important aspect of the nature of light is *frequency*. The frequency of a wave is its rate of oscillation and is measured in hertz, the SI unit of frequency, where one hertz is equal to one oscillation per second. Light usually has a spectrum of frequencies which sum together to form the resultant wave. Different frequencies undergo different angles of refraction.

A wave consists of successive troughs and crests, and the distance between two adjacent crests or troughs is called the *wavelength*. Waves of the electromagnetic spectrum vary in size, from very long radio waves the size of buildings to very short gamma rays smaller than atom nuclei. Frequency is inversely proportional to wavelength, according to the equation:

velocity (the speed of light), wavelength, and frequency. When considered as particles, they are known as photons, and each has an energy related to the frequency of the wave given by Planck's relation E = hν, where E is the energy of the photon, h = 6.626 × 10^{-34} J·s is Planck's constant, and υ is the frequency of the wave.

One rule is always obeyed regardless of the circumstances: EM radiation in a vacuum always travels at the speed of light, relative to the observer, regardless of the observer's velocity. (This observation led to Albert Einstein's development of the theory of special relativity.)

In a medium (other than vacuum), velocity factor or refractive index are considered, depending on frequency and application. Both of these are ratios of the speed in a medium to speed in a vacuum.

Electromagnetic Spectrum

Generally, EM radiation is classified by wavelength into electrical energy, radio, microwave, infrared, the visible region we perceive as light, ultraviolet, X-rays and gamma rays.

The behaviour of EM radiation depends on its wavelength. Higher frequencies have shorter wavelengths, and lower frequencies have longer wavelengths. When EM radiation interacts with single atoms and molecules, its behavior depends on the amount of energy per quantum it carries. Spectroscopy can detect a much wider region of the EM spectrum than the visible range of 400 nm to 700 nm. A common laboratory spectroscope can detect wavelengths from 2 nm to 2500 nm. Detailed information about the physical properties of objects, gases, or even stars can be obtained from this type of device. It is widely used in astrophysics. For example, hydrogen atoms emit radio waves of wavelength 21.12 cm.

Light

The EM radiation with a wavelength between approximately 400 nm and 700 nm is detected by the human eye and perceived as visible light. Other wavelengths, especially nearby infrared (longer than 700 nm) and ultraviolet (shorter than 400 nm) are also sometimes referred to as light, especially when the visibility to humans is not relevant.

If radiation having a frequency in the visible region of the EM spectrum reflects off of an object, say, a bowl of fruit, and then strikes our eyes, this results in our visual perception of the scene. Our brain's visual system processes the multitude of reflected frequencies into different shades and hues, and through this not-entirely-understood psycho-physical phenomenon, most people perceive a bowl of fruit.

At most wavelengths, however, the information carried by electromagnetic radiation is not directly detected by human senses. Natural sources produce EM radiation across the spectrum, and our technology can also manipulate a broad range of wavelengths. Optical fiber transmits light which, although not suitable for direct viewing, can carry data that can be translated into sound or an image. The coding used in such data is similar to that used with radio waves.

Radio Waves

Radio waves can be made to carry information by varying a combination of the amplitude, frequency and phase of the wave within a frequency band.

When EM radiation impinges upon a conductor, it couples to the conductor, travels along it, and induces an electric current on the surface of that conductor by exciting the electrons of the conducting material. This effect (the skin effect) is used in antennas. EM radiation may also cause certain molecules to absorb energy and thus to heat up; this is exploited in microwave ovens.

Derivation

Electromagnetic waves as a general phenomenon were predicted by the classical laws of electricity and magnetism, known as Maxwell's equations. If you inspect Maxwell's equations without sources (charges or currents) then you will find that, along with the possibility of nothing happening, the theory will also admit nontrivial solutions of changing electric and magnetic fields. Beginning with Maxwell's equations for free space:

$$\nabla . E = 0 \quad (7.1)$$

$$\nabla \times \mathrm{E} = \frac{\partial \mathrm{B}}{\partial t} \quad (7.2)$$

$$\nabla . \mathrm{B} = 0 \quad (7.3)$$

$$\nabla \times \mathrm{B} = \mu_0 \varepsilon_0 \frac{\partial \mathrm{E}}{\partial t} \quad (7.4)$$

where

∇ is a vector differential operator.

One solution

$\mathrm{E} = \mathrm{B} = 0$

is trivial.

To see the more interesting one, we utilize vector identities, which work for any vector, as follows:

$$\nabla \times (\nabla \times \mathrm{A}) = \nabla(\nabla . \mathrm{A}) - \nabla^2 \mathrm{A}$$

To see how we can use this take the curl of eq. (7.2):

$$\nabla \times (\nabla \times \mathrm{E}) = \nabla \times \left(-\frac{\partial B}{\partial t} \right) \quad (7.5)$$

Evaluating the left hand side:

$$\nabla \times (\nabla \times E) = \nabla(\nabla . E) - \nabla^2 E = -\nabla^2 E \tag{7.6}$$

where we simplified the above by using equation (1). Evaluate the right hand side:

$$\nabla \times \left(\frac{\partial B}{\partial t} \right) = -\frac{\partial}{\partial t} (\nabla \times B) = -\mu_0 \varepsilon_0 \frac{\partial^2 E}{\partial t^2} \tag{7.7}$$

Equations (7.6) and (7.7) are equal, so this results in a vector-valued differential equation for the electric field, namely

$$\nabla^2 E = \mu_0 \varepsilon_0 \frac{\partial^2 E}{\partial t^2}$$

Applying a similar pattern results in similar differential equation for the magnetic field:

$$\nabla^2 B = \mu_0 \varepsilon_0 \frac{\partial^2 B}{\partial t^2}$$

These differential equations are equivalent to the wave equation:

$$\nabla^2 f = \frac{1}{c0^2} \frac{\partial^2 f}{\partial t^2}$$

where

c_0 is the speed of the wave in free space and

f describes a displacement

Or more simply

$$\Box^2 f = 0$$

where $\Box^2$ is d' Alembertian:

$$\Box^2 = \nabla^2 - \frac{1}{c_0^{\ 2}}\frac{\partial^2}{\partial t^2} = \frac{\partial^2}{\partial x^2} + \frac{\partial^2}{\partial y^2} + \frac{\partial^2}{\partial z^2} - \frac{1}{c_0^{\ 2}}\frac{\partial^2}{\partial t_2}$$

Notice that in the case of the electric and magnetic fields, the speed is:

$$c_0 = \frac{1}{\sqrt{\mu_0 \varepsilon_0}}$$

Which, as it turns out, is the speed of light in free space. Maxwell's equations have unified the permittivity of free space ε_0, the permeability of free space μ_0, and the speed of light itself, c_0. Before this derivation it was not known that there was such a strong relationship between light and electricity and magnetism.

But these are only two equations and we started with four, so there is still more information pertaining to these waves hidden within Maxwell's equations. Let's consider a generic vector wave for the electric field.

$$E = E_0 f\,(\hat{k}.x - c_0 t)$$

Here E_0 is the constant amplitude, f is any second differentiable function, $\hat{k}$ is a unit vector in the direction of propagation, and x is a position vector. We observe that

$$f\,(\hat{k}.x - c_0 t)$$

is a generic solution to the wave equation. In other words

$$\nabla^2 f(\hat{k}.x - c_0 t) = \frac{1}{c_0^{\ 2}}\frac{\partial^2}{\partial t_2} f\,(\hat{k}.x - c_0 t)$$

for a generic wave traveling in the $\hat{k}$ direction.

This form will satisfy the wave equation, but will it satisfy all of Maxwell's equations, and with what corresponding magnetic field?

$$\nabla.E = \hat{k}.E_0 f\,(k.x - c_0 t) = 0$$

$$E = \hat{k} = 0$$

The first of Maxwell's equations implies that electric field is orthogonal to the direction the wave propagates.

$$\nabla \times E = \hat{k} \times E_0 f\,(\hat{k}.x - c_0 t) = -\frac{\partial B}{\partial t}$$

$$B = \frac{1}{c_0}\hat{k} \times E$$

The second of Maxwell's equations yields the magnetic field. The remaining equations will be satisfied by this choice of **E, B**.

Not only are the electric and magnetic field waves traveling at the speed of light, but they have a special restricted orientation and proportional magnitudes, $E_0 = c_0 B_0$, which can be seen immediately from the Poynting vector. The electric field, magnetic field, and direction of wave propagation are all orthogonal, and the wave propagates in the same direction as **E** x **B**.

From the viewpoint of an electromagnetic wave traveling forward, the electric field might be oscillating up and down, while the magnetic field oscillates right and left; but this picture can be rotated with the electric field oscillating right and left and the magnetic field oscillating down and up. This is a different solution that is traveling in the same direction. This arbitrariness in the orientation with respect to propagation direction is known as polarization.

ELECTROMAGNETIC SPECTRUM

The electromagnetic spectrum is the range of all possible electromagnetic radiation frequencies The "electromagnetic spectrum" (usually just spectrum) of an object is the characteristic distribution of electromagnetic radiation from that particular object.

The electromagnetic spectrum extends from below the frequencies used for modern radio (at the long-wavelength end) through gamma radiation (at the short-wavelength end), covering wavelengths from thousands of kilometers down to a fraction the size of an atom. It is thought that the short wavelength limit is in the vicinity of the Planck length, and the long wavelength limit is the size of the universe itself (see physical cosmology), although in principle the spectrum is infinite and continuous.

Range of the Spectrum

The EM waves are typically described by any of the following three physical properties: the frequency, f, and wavelength λ, and photon energy E. Frequencies range from about a million billion Hertz (gamma rays) down to a few Hertz (radio waves). Wavelength is inversely proportional to the wave frequency, so gamma rays have very short wavelengths that are fractions of the size of atoms, whereas radio wavelengths can be as long as a several thousand kilometers. Photon energy is directly proportional to the wave frequency, so gamma rays have the highest energy around a mega electron volt and radio waves have very low energy around femto electron volts (femto = 10-15). These relations are illustrated by the following equations:

$$f = \frac{c}{1} \text{ or } f = \frac{E}{h} \text{ or } E = \frac{hc}{1},$$

where:

c = 299,792,458 m/s (speed of light in vacuum); and

h = 6.62606896(33) × 10-34 J·s (Planck's constant).

Whenever light waves (and other electromagnetic waves) exist in a medium (matter), their wavelength is decreased. Wavelengths of electromagnetic radiation, no matter what medium they are traveling through, are usually quoted in terms of the vacuum wavelength , although this is not always explicitly stated. Generally, EM radiation is classified by coiled wavelength into radio wave, microwave, infrared, the visible region we perceive as light, ultraviolet, X-rays and gamma rays.

The behaviour of EM radiation depends on its wavelength. When EM radiation interacts with single atoms and molecules, its behavior also depends on the amount of energy per quantum (photon) it carries. Electromagnetic radiation can be divided into octaves — as sound waves.

Spectroscopy can detect a much wider region of the EM spectrum than the visible range of 400 nm to 700 nm. A common laboratory spectroscope can detect wavelengths from 2 nm to 2500 nm. Detailed information about the physical properties of objects, gases, or even stars can be obtained from this type of device. It is widely used in astrophysics. For example, many hydrogen atoms emit a radio wave photon which has a wavelength of 21.12 cm. Also, frequencies of 30 Hz and below can be produced by and are important in the study of certain stellar nebulae and frequencies as high as 2.9×1027 Hz have been detected from astrophysical sources.

Types of Radiation

While the classification scheme is generally accurate, in reality there is often some overlap between neighboring types of electromagnetic energy. For example, SLF radio waves at 60 Hz may be received and studied by astronomers, or may be ducted along wires as electric power. Also, some low-energy gamma rays actually have a longer wavelength than some high-energy X-rays. This is possible because "gamma ray" is the name given to the photons generated from nuclear decay or other nuclear and subnuclear processes, whereas X-rays on the other hand are generated by electronic transitions involving highly energetic inner electrons. Therefore the distinction between gamma ray and X-ray is related to the radiation source rather than the radiation wavelength. Generally, nuclear transitions are much more energetic than electronic transitions, so usually, gamma-rays are more energetic than X-rays. However, there are a few low-energy nuclear transitions (e.g. the 14.4 keV nuclear transition of Fe-57) that produce gamma rays that are less energetic than some of the higher energy X-rays.

Radio Frequency

Radio waves generally are utilized by antennas of appropriate size (according to the principle of resonance), with wavelengths ranging from hundreds of meters to about one millimeter. They are used for transmission of data, via modulation. Television, mobile phones, wireless networking and amateur radio all use radio waves.

Radio waves can be made to carry information by varying a combination of the amplitude, frequency and phase of the wave within a frequency band and the use of the radio spectrum is regulated by many governments through frequency allocation. When EM radiation impinges upon a conductor, it couples to the conductor, travels along it, and induces an electric current on the surface of that conductor by exciting the electrons of the conducting material. This effect (the skin effect) is used in antennas. EM radiation may also cause certain molecules to absorb energy and thus to heat up, thus causing thermal effects and sometimes burns; this is exploited in microwave ovens.

Microwaves

The super high frequency (SHF) and extremely high frequency (EHF) of microwaves come next up the frequency scale. Microwaves are waves which are typically short enough to employ tubular metal waveguides of reasonable diameter. Microwave energy is produced with klystron and magnetron tubes, and with solid state diodes such as Gunn and IMPATT devices. Microwaves are absorbed by molecules that have a dipole moment in liquids. In a microwave oven, this effect is used to heat food. Low-intensity microwave radiation is used in Wi-Fi, although this is at intensity levels unable to cause thermal heating.

Volumetric heating, as used by microwaves, transfer energy through the material electro-magnetically, not as a thermal heat flux. The benefit of this is a more uniform heating and reduced heating time; microwaves can heat material in less than 1% of the time of conventional heating methods.

When active, the average microwave oven is powerful enough to cause interference at close range with poorly shielded electromagnetic fields such as those found in mobile medical devices and cheap consumer electronics.

Terahertz Radiation

Terahertz radiation is a region of the spectrum between far infrared and microwaves. Until recently, the range was rarely studied and few sources existed for microwave energy at the high end of the band (sub-millimetre waves or so-called terahertz waves), but applications such as imaging and communications are now appearing. Scientists are also looking to apply terahertz technology in the armed forces, where high frequency waves might be directed at enemy troops to incapacitate their electronic equipment.

Infrared Radiation

The infrared part of the electromagnetic spectrum covers the range from roughly 300 GHz (1 mm) to 400 THz (750 nm). It can be divided into three parts:

- *Far-infrared,* from 300 GHz (1 mm) to 30 THz (10 μm). The lower part of this range may also be called microwaves. This radiation is typically absorbed by so-called rotational modes in gas-phase molecules, by molecular motions in liquids, and by phonons in solids. The water in the Earth's atmosphere absorbs so strongly in this range that it renders the atmosphere effectively opaque. However, there are certain wavelength ranges ("windows") within the opaque range which allow partial transmission, and can be used for astronomy. The wavelength range from approximately 200 μm up to a few mm is often referred to as "sub-millimetre" in astronomy, reserving far infrared for wavelengths below 200 μm.
- *Mid-infrared,* from 30 to 120 THz (10 to 2.5 μm). Hot objects (black-body radiators) can radiate strongly in this range. It is absorbed by molecular vibrations, where the different atoms in a molecule vibrate around their equilibrium positions. This range is sometimes called the fingerprint region since the mid-infrared absorption spectrum of a compound is very specific for that compound.
- *Near-infrared,* from 120 to 400 THz (2,500 to 750 μm). Physical processes that are relevant for this range are similar to those for visible light.

Visible Radiation (Light)

Above infrared in frequency comes visible light. This is the range in which the sun and stars similar to it emit most of their radiation. It is probably not a coincidence that the human eye is sensitive to the wavelengths that the sun emits most strongly. Visible light (and near-infrared light) is typically absorbed and emitted by electrons in molecules and atoms that move from one energy level to another. The light we see with our eyes is really a very small portion of the electromagnetic spectrum. A rainbow shows the optical (visible) part of the electromagnetic spectrum; infrared (if you could see it) would be located just beyond the red side of the rainbow with ultraviolet appearing just beyond the violet end.

EM radiation with a wavelength between 380 nm and 760 nm is detected by the human eye and perceived as visible light. Other wavelengths, especially near infrared (longer than 760 nm) and ultraviolet (shorter than 380 nm) are also sometimes referred to as light, especially when the visibility to humans is not relevant.

If radiation having a frequency in the visible region of the EM spectrum reflects off of an object, say, a bowl of fruit, and then strikes our eyes, this results in our visual perception of the scene. Our brain's visual system processes the multitude of reflected frequencies into different shades and hues, and through this not-entirely-understood psycho-physical phenomenon, most people perceive a bowl of fruit.

At most wavelengths, however, the information carried by electromagnetic radiation is not directly detected by human senses. Natural sources produce EM radiation across the spectrum, and our technology can also manipulate a broad range of wavelengths. Optical fiber transmits light which, although not suitable for direct viewing, can carry data that can be translated into sound or an image. The coding used in such data is similar to that used with radio waves.

Ultraviolet Light

Next in frequency comes ultraviolet (UV). This is radiation whose wavelength is shorter than the violet end of the visible spectrum, and longer than that of an x-ray. Being very energetic, UV can break chemical bonds, making molecules unusually reactive or ionizing them, in general changing their mutual behavior. Sunburn, for example, is caused by the disruptive effects of UV radiation on skin cells, which can even cause skin cancer, if the radiation irreparably damages the complex DNA molecules in the cells (UV radiation is a proven mutagen). The Sun emits a large amount of UV radiation, which could quickly turn Earth into a barren desert; however, most of it is absorbed by the atmosphere's ozone layer before reaching the surface.

X-rays

After UV come X-rays. Hard X-rays have shorter wavelengths than soft X-rays. X-rays are used for seeing through some things and not others, as well as for high-energy physics and astronomy. Neutron stars and accretion disks around black holes emit X-rays, which enable us to study them.

X-rays will pass through most substances, and this makes them useful in medicine and industry. X-rays are given off by stars, and strongly by some types of nebulae. An X-ray machine works by firing a beam of electrons at a "target". If the electrons are fired with enough energy, X-rays will be produced.

Gamma Rays

After hard X-rays come gamma rays, which were discovered by Paul Villard in 1900. These are the most energetic photons having no defined lower limit to their wavelength. They are useful to astronomers in the study of high energy objects or regions and find a use with physicists thanks to their penetrative ability and their production from radioisotopes. The wavelength of gamma rays can be measured with high accuracy by means of Compton scattering.

Note that there are no precisely defined boundaries between the bands of the electromagnetic spectrum. Radiation of some types have a mixture of the properties of those in two regions of the spectrum. For example, red light resembles infrared radiation in that it can resonate some chemical bonds.

CHAPTER 8

TWO-PHOTON ABSORPTION

Two photon absorption (TPA) is the simultaneous absorption of two photons of identical or different frequencies in order to excite a molecule from one state (usually the ground state) to a higher energy electronic state. The energy difference between the involved lower and upper states of the molecule is equal to the sum of the energies of the two photons. Two-photon absorption is many orders of magnitude weaker than linear absorption and is therefore not an everyday phenomenon. It differs from linear absorption in that the strength of absorption depends on the square of the light intensity, thus it is a nonlinear optical process.

Background

The phenomenon was originally predicted by Maria Göppert-Mayer in 1931 in her doctoral dissertation. Thirty years later, the invention of the laser permitted the first experimental verification of the TPA when two-photon-excited fluorescence was detected in a europium-doped crystal and then observed in a vapour (cesium) in 1962 by Isaac Abella.

TPA is a third-order nonlinear optical process. In particular, the imaginary part of the third-order nonlinear susceptibility is related to the extent of TPA in a given molecule. The selection rules for TPA are therefore different than for one-photon absorption (OPA), which is dependent on the first-order susceptibility. For example, in a centrosymmetric molecule, one- and two-photon allowed transitions are mutually exclusive. In quantum mechanical terms, this difference results from the need to conserve spin. Since photons have spin of ±1, one-photon absorption requires excitation to involve an electron changing its molecular orbital to one with a spin different by ±1. Two-photon absorption requires a change of +2, 0, or -2.

The third order can be rationalized by considering that a second order process creates a polarization with the doubled frequency. In the third order, by difference frequency generation the original frequency can be generated again. Depending on the phase between the generated polarization and the original electric field this leads to the Kerr effect or to the *two-photon absorption*. In second harmonic generation this difference in frequency generation is a separated process in a cascade, so that the energy of the fundamental frequency can also be absorbed. This may be better called three photon absorption. In the next paragraph resonant two photon absorption via separate one-photon transitions is mentioned, where the absorption alone is a first order process and any fluorescence from the final state of the second transition will be of second order; this means it will rise as the square of the incoming intensity. The virtual state argument is quite orthogonal to the anharmonic oscillator argument. It states for example that in a semiconductor absorption at high energies is impossible, if two photons cannot bridge the band gap. So a many materials can be used for the Kerr effect that do not show any absorption and thus have a high damage threshold.

Two-photon absorption can be measured by several techniques. Two of them are two-photon excited fluorescence (TPEF) and nonlinear transmission (NLT). Pulsed lasers are most often used because TPA is a third-order nonlinear optical process, and therefore is most efficient at very high intensities. Phenomenologically, this can be thought of as the third term in a conventional anharmonic oscillator model for depicting vibrational behavior of molecules. Another view is to think of light as photons. In nonresonant TPA two photons combine to bridge an energy gap larger than the energies of each photon individually. If there were an intermediate state in the gap, this could happen via two separate one-photon transitions in a process described as "resonant TPA", "sequential TPA", or "1+1 absorption". In nonresonant TPA the transition occurs without the presence of the intermediate state. This can be viewed as being due to a "virtual" state created by the interaction of the photons with the molecule.

The "nonlinear" in the description of this process means that the strength of the interaction increases faster than linearly with the electric field of the light. In fact, under ideal conditions the rate of TPA is proportional to the square of the field intensity. This dependence can be derived quantum mechanically, but is intuitively obvious when one considers that it requires two photons to coincide in time and space. This requirement for high light intensity means that lasers are required to study TPA phenomena. Further, in order to understand the TPA spectrum, monochromatic light is also desired in order to measure the TPA cross section at different wavelengths. Hence, tunable pulsed lasers (such as frequency-doubled Nd:YAG-pumped OPOs and OPAs) are the choice of excitation.

Measurements

Absorption Rate

The Beer's law for OPA

$$I(x) = I_0 e^{-\alpha cx}$$

changes to

$$I(x) = \frac{I_0}{1 + \beta cx I_0}$$

for TPA with light intensity as a function of path length or cross section x as a function of concentration c and the initial light intensity I_0. The absorption coefficient a now becomes the TPA *cross section* ß. (Note that there is some confusion over the term ß in non-linear optics, since it is sometimes used to describe the second-order polarizability, and occasionally for the molecular two-photon cross-section. More often however, is it used to describe the bulk 2-photon optical density of a sample. The letter d or s is more often used to denote the molecular two-photon cross-section.)

Units of Cross-section

The molecular two-photon cross-section is usually quoted in the units of Göppert-Mayer (GM) (after its discoverer, Nobel laureate Maria Göppert-Mayer), where $^{-1}$GM is 10^{-50}cm^4.s.photon^{-1}molecules^{-1}. Considering the reason for these units, one can see that it results from the product of two areas (one for each photon, each in cm^2) and a time (within which the two photons must arrive to be able to act together). The large scaling factor is introduced in order that 2-photon absorption cross-sections of common dyes will have convenient values.

Development of the Field and Potential Applications

Until the early 1980s, TPA was used as a spectroscopic tool. Scientists compared the OPA and TPA spectra of different organic molecules and obtained several fundamental structure property relationships. However, in late 1980s, applications were started to be developed. Peter

Rentzepis suggested applications in 3D optical data storage. Watt Webb suggested microscopy and imaging. Other applications such as 3D microfabrication and optical power limiting were also suggested.

Microfabrication and Lithography

One of the most distinguishing features of TPA is that the rate of absorption of light by a molecule depends on the square of the light's intensity. This is different than OPA, where the rate of absorption is linear with respect to input intensity. As a result of this dependence, if material is cut with a high power laser beam, the rate of material removal decreases very sharply from the center of the beam to its periphery. Because of this, the "pit" created is sharper and better resolved than if the same size pit were created using normal absorption.

3D Photopolymerization

In 3D microfabrication, a block of gel containing monomers and a 2-photon active photoinitiator is prepared as a raw material. Application of a focused laser to the block results in polymerization only at the focal spot of the laser, where the intensity of the absorbed light is highest. The shape of an object can therefore be traced out by the laser, and then the excess gel can be washed away to leave the traced solid.

Imaging

The human body is not transparent to visible wavelengths. Hence, one photon imaging using fluorescent dyes is not very efficient. If the same dye had good two-photon absorption, then the corresponding excitation would occur at approximately two times the wavelength at which one-photon excitation would have occurred. As a result, it is possible to use excitation in the far infrared region where the human body shows good transparency. It is sometimes said, incorrectly, that Rayleigh scattering is relevant to imaging techniques such as two-photon. According to

Rayleigh's scattering law, the amount of scattering is proportional to $1/\lambda^4$, where λ is the wavelength. As a result, if the wavelength is increased by a factor of 2, the Rayleigh scattering is reduced by a factor of 16. However, Rayleigh scattering only takes place when scattering particles are much smaller than the wavelength of light (the sky is blue because air molecules scatter blue light much more than red light). When particles are larger, scattering increases approximately linearly with wavelength: hence clouds are white since they contain water droplets. This form of scatter is known as Mie scattering and is what occurs in biological tissues. So, although longer wavelengths do scatter less in biological tissues, the difference is not as dramatic as Rayleigh's law would predict.

Optical Power Limiting

Another area of research is optical power limiting. In a material with a strong nonlinear effect, the absorption of light increases with intensity such that beyond a certain input intensity the output intensity approaches a constant value. Such a material can be used to limit the amount of optical power entering a system. This can be used to protect expensive or sensitive equipment such as sensors, can be used in protective goggles, or can be used to control noise in laser beams.

Photodynamic Therapy

Photodynamic therapy (PDT) is a method for treating cancer. In this technique, an organic molecule with a good triplet quantum yield is excited so that the triplet state of this molecule interacts with oxygen. The ground state of oxygen has triplet character. This leads to triplet-triplet annihilation, which gives rise to singlet oxygen, which in turn attacks cancerous cells. However, using TPA materials, the window for excitation can be extended into the infrared region, thereby making the process more viable to be used on the human body.

Optical Data Storage

The ability of two-photon excitation to address molecules deep within a sample without affecting other areas makes it possible to store and retrieve information in the volume of a substance rather than only on a surface as is done on the DVD. Therefore, 3D optical data storage has the possibility to provide media that contain terabyte-level data capacities on a single disc.

TPA Compounds

To some extent, linear and 2-photon absorption strengths are linked. Therefore, the first compounds to be studied (and many that are still studied and used in e.g. 2-photon microscopy) were standard dyes. In particular, laser dyes were used, since these have good photostability characteristics. However, these dyes tend to have 2-photon cross-sections of the order of 0.1-10 GM, much less than is required to allow simple experiments.

It was not until the 1990s that rational design principles for the construction of two-photon-absorbing molecules began to be developed, in response to a need from imaging and data storage technologies, and aided by the rapid increases in computer power that allowed quantum calculations to be made. The accurate quantum mechanical analysis of two-photon absorbance is orders of magnitude more computationally intensive than that of one-photon absorbance, requiring highly correlated calculations at very high levels of theory.

The most important features of strongly TPA molecules were found to be a long conjugation system (analogous to a large antenna) and substitution by strong donor and acceptor groups (which can be thought of as inducing nonlinearity in the system and increasing the potential for charge-transfer). Therefore, many push-pull olefins exhibit high TPA transitions, up to several thousand GM. It is also found that compounds with a real intermediate energy level close to the "virtual" energy level can have large 2-photon cross-sections as a result of resonance enhancement.

Compounds with interesting TPA properties also include various porphyrin derivatives, conjugated polymers and even dendrimers. In one study a diradical resonance contribution for the compound depicted below was also linked to efficient TPA. The TPA wavelength for this compound is 1425 nanometer with observed TPA cross section of 424 GM.

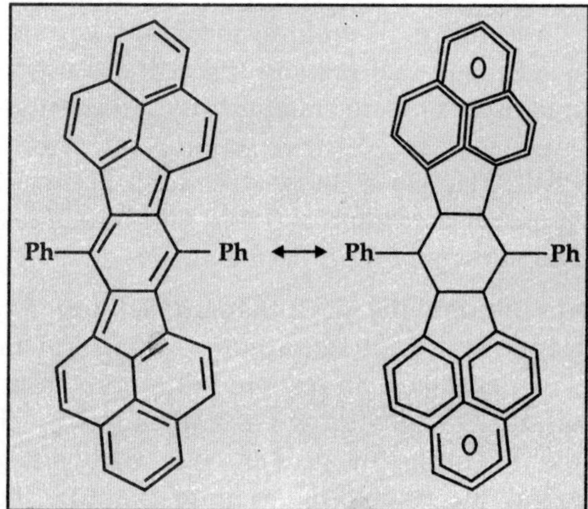

Fig. 8.1

VISIBLE LIGHT AND X-RAYS

Photons are absorbed in a solid through a process known as the photoelectric effect whereby a photon interacts with an electron by giving all its energy to the electron. In effect, the photon disappears and energy is transferred to an electron. The photoelectric process is responsible for the absorption of visible-light with energies of 2 to 3 eV as well as for x-rays with energies a thousand times greater than that of visible light.

The absorption of photons in solids depends on the number of electrons that can accept a transfer of energy from the photon. Not all electrons meet this criterion. Some electrons are so tightly bound to their orbital around the atomic nucleus that the photon energy is unable to break the bonds. Other electrons are involved in the bonding between atoms and cannot be freed by the energy of the incident photons. For example, ordinary window glass is transparent to visible photons, but is strongly absorbent to ultraviolet radiation where the photons have energies only a few times greater than that of visible light.

Photon absorption in the visible region of the spectrum depends on the atomic arrangement of the atoms and their bonding. Pure silicon (Si) is strongly absorbing but silicon combined with oxygen is transparent. For the energetic photons in the x-ray regime, photon absorption is much easier to predict and is independent of the details of atomic arrangement. It depends primarily on the electron concentration per unit volume. Since the concentration of atoms per unit volume only differ by factors of 2 or 3 from each other, the electron concentration in two materials can be estimated from the atomic number, Z. Lead (Z=82) absorbs x-rays much more efficiently than aluminum (Z=13) and consequently is used in shielding around x-ray apparatus. X-ray absorption does depend on the energy of the x-rays and decreases with increasing x-ray energy, E. Absorption decreases nearly proportional to the cube of the energy (i.e. absorption proportional to (1/E3).

In spite of the different material dependencies in the description of visible light and x-ray absorption, the mathematical description is the same for the two: the exponential decay law. This decay law is due to the fact that each successive interval of equal thickness of a sample absorbs an equal fraction of the photons passing through it. Each successive interval, however, receives few photons than the preceding interval because the number of photons decreases with depth. For example, if the first layer thickness absorbs one-half of the photons, then only half of the original number of photons are incident on the second layer thickness which absorbs one-half of these, i.e. on quarter of the original number of photons incident on the sample surface.

Mechanism for the Absorption of Light

The formalism for the absorption of light in the simplest case starts with a beam of light of intensity I of photons that has just penetrated into a material with a uniform concentration of absorbing pigments. We ignore reflection and refraction. The photons penetrate into the material and are absorbed at different depths (Fig. 8.2).

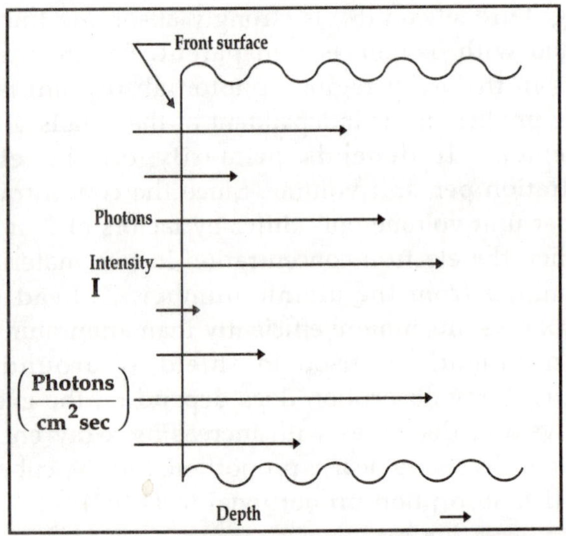

Fig. 8.2: Schematic of light rays penetrating to different depths into a photon absorbing material.

The mechanism for absorption is that a photon transfers all its energy to an electron in the absorbing material. The photon is "lost" from the light beam as it is absorbed in a single event. The electron is excited by the gain in energy to a higher energy state in the electron configuration around the atom. This single-event transfer of photon energy to an electron and the disappearance of the photon is closely related to the photoelectric effect where the electron is ejected from the material in a single event. The electron is ejected from the atom by the absorption of an x-ray. This loss of an electron leaves the atom in an excited state.

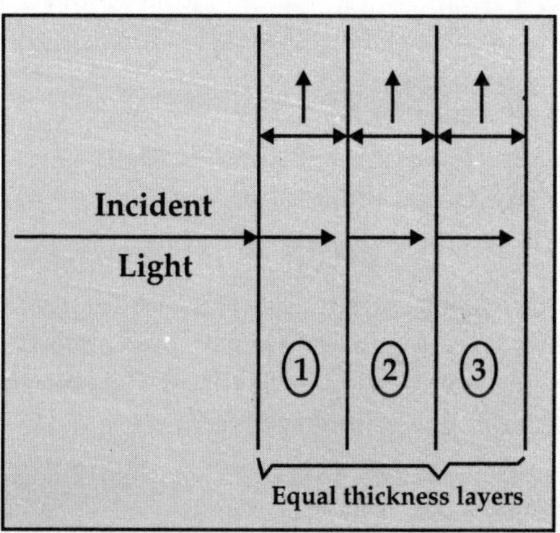

Fig. 8.3: Light incident on a light absorbing material that is represented by a stack of equivalent thickness, t, layers. The length of the arrows decrease as the light penetrate showing the decreasing intensity of light.

The decrease in intensity of the light or photon flux as the beam penetrates into the material can be visualized if we separate the absorbing material into a set of thin, equal thickness slices of thickness *t* (Fig. 8.3). The material is homogeneous and the thickness t is chosen so that only a small fraction of the light is absorbed in passing through the layer. The symbol A is used to denote the fraction absorbed. The fraction of light absorbed in the first layer is (1-A) times the incident intensity. For example, if the fraction of light absorbed per layer is 10% of the incident flux, then for 100 photons incident on the first layer, AI=0.1 × 100 or 10 will be absorbed and 90 transmitted to the second layer.

The light intensity incident on the second layer is (1-A)I and A(1-A)I will be absorbed on the second layer. The amount transmitted through the second layer is (1-A)2I. To continue our example, with 90 photons incident on the second layer, A × 90 = 0.1 × 90 = 9 photons will be absorbed and 81 transmitted. That is:

Layer 1, I photons incident, AI=0.1 × 100 = 10 absorbed and

(1-A) I=(1-.1)100=90 transmitted

Layer 2, (1-A) I=90 photons incident,
A(1-A)I=0.1(0.9)100=9 absorbed and

(1-A)(1-A)I=(0.9)(0.9)100=81 are transmitted.

The process continues with fewer photons being transmitted into layer 3, layer 4, and so on, with fewer photons being absorbed in each successive layer although the fraction absorbed A remains constant.

For n layers, the total absorption TA in all layers absorbed fraction A will be

$$TA=A + A(1-A) + A(1-A)^2 + \dots + A(1-A)\ n-1 \qquad (1)$$

The first time in equation 1 represents the absorption by the first layer. The second term represents the absorption by the second layer, and so forth.

The total transmitted light TT though n layer is

$$TT = (1 - A)^2 \qquad (2)$$

Equation 1 is more conveniently written using an exponential function. Using an absorption coefficient alpha, the intensity of light at depth t, I(t) for an incident intensity I0 is given by

$$I(t) = I_0 \exp(-\alpha t) \qquad (3)$$

where alpha is given as a fraction per distance such as 0.2 per cm. Figure 8.4 shows the intensity of light as a function of penetration distance into materials with two different absorption coefficients. The exponential decrease is illustrated on a linear scale by the curves in Fig. 8.4. An increase in alpha leads to a decrease in penetration. At the depth where the product alpha t equals unity, the intensity has dropped 1/e of a factor of 0.37 times the incident intensity.

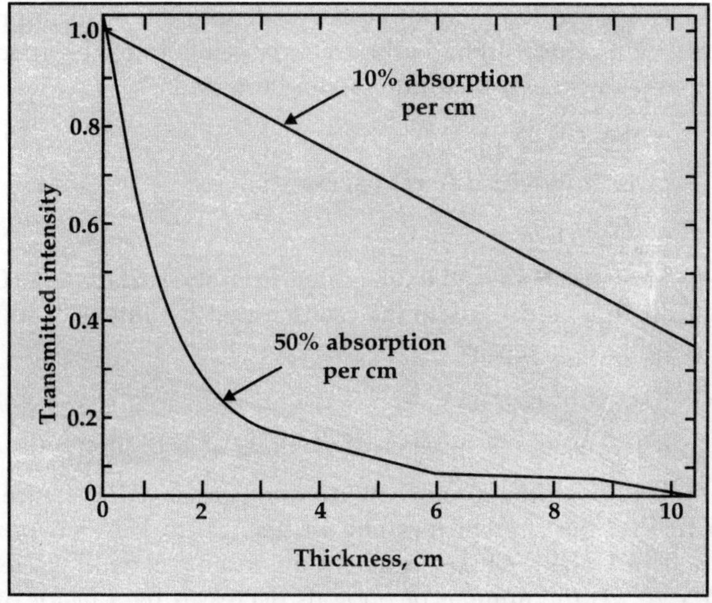

Fig. 8.4: The intensity of light transmitted through an absorbing material as a function of thickness d in cm for (1) 10% absorbance and (2) 50% absorbance per cm.

The relations rely on the fact that the fraction of light which is absorbed is independent of the intensity of the incident light. For a given wavelength of light, each successive equi-thickness of the medium aborbs an equal fraction of light passing through it, but the amount of light it receives is sucessively decreasing. Hence, the intensity of light passing though a medium will decrease with depth. The exponential attenuation coefficients apply for x-rays as well as photons in the visible portion of the electromagnetic spectrum. The exponential relation is derived in the following:

The number of deltaI of photons which are absorbed in a given thickness deltax is proportional to the incident intensity I or the number I incident on it:

$$-\Delta I \textit{ proportional to } I\Delta x \quad (4)$$

If we use the symbol alpha to denote the absorption coefficient where alpha is the fraction of photons absorbed per unit thickness, then Eqn. 4 can be written:

$$-\Delta I = \alpha I \Delta x \quad (5)$$

or in standard differential notation as:

$$-dI = \alpha I\, dx \quad (6)$$

This equation can be integrated from the surface at X=0 to a depth x, and leads to the relation that the intensity I(x) at depth x is given by

$$I(x) = I_0 \, exp(-\alpha x) \quad (7)$$

where I0 is the number of photons at x=0, I0=I(x=0).

The absorption coefficient is a number per unit distance, such as 0.1 per cm which can be written 0.1 cm-1. At a depth x, where alpha x equals unity (in this example, x=10 cm as 0.1 x 10 =1) the number of photons decreases by a factor of 1/e or 0.37 (the value of e=2.718).

For small values of alpha x, for example alpha x =0.01, Eqn (7) can be written as a linear expansion:

$$I(x) = I_0 \,(1 - \alpha x) \quad (8)$$

For x-rays, photoelectric absorption (where the x-ray gives up its energy in a single interaction with an electron) is the major cause of attenuation of the photons penetrating the material.

The intensity I of x-rays transmitted through a thin foil of material for an incident intensity I_0 follows the exponential attenuation relation of equation (3) with a change in the nomenclature from alpha to mu to follow the convention of in standard x-ray usage:

$$I = I_0 \exp(-\mu x) = I_0 \exp[(-\mu/\rho)\ \ x] \tag{9}$$

where rho is the density of the element (g/cm^3), mu is the linear attenuation coefficient, and mu/rho is the mass attenuation coefficient given in cm^2/g. Note that x-ray absorption is not influenced by the crystal structure of the pigment, but only by the number of atoms/cm^3 and the thickness of the pigment layer.

X-Radiography

An X-ray image of the internal structure of an object is called a Radiograph. The object is called the Specimen. The Specimen is placed between an X-ray generator, called the Source, and the recording medium, called the Target. In medical or art museum X-Radiography, the recording medium is special photographic film that is sensitive to X-rays. X-ray film becomes exposed wherever X-rays strike the Target. More X-rays at the Target cause greater exposure of the film and fewer X-rays at the Target cause less exposure of the film. Where the Specimen blocks the X-rays by absorbing them, thie Target does not get exposed. On the other hand, where the Specimen does not absorb the X-rays, they pass through the Specimen and the Target becomes very exposed. In most cases, the amount of exposure at the Target can vary continuously from fully exposed to not exposed at all, depending on the internal structure and composition of the Specimen. Figure 8.5 is a schematic view of the procedure.

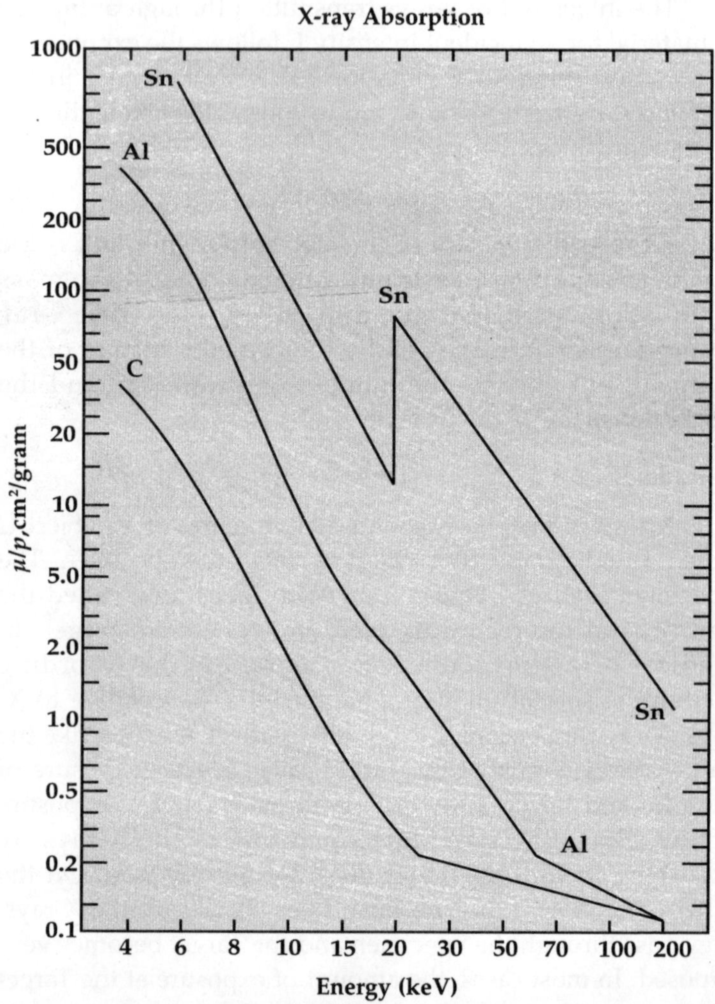

Fig. 8.5: The x-ray mass absorption coefficient (mu/rho) in cm2/ gram versus x-ray energy in kiloelectron volts (keV) for carbon (C), aluminum (Al), and tin (Sn). The sharp increase in absorption for Sn at 29.2 keV corresponds in energy to the K-shell, 1s binding energy (taken from Feldman and Mayer, Fundamentals of Surface and Thin Film Analysis.

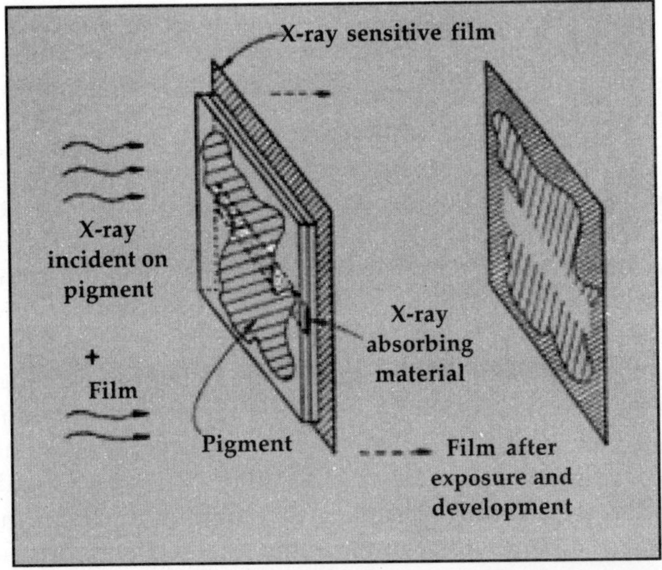

Fig. 8.6

Upon development of the X-ray film, the differences in exposure produce differences in contranst on the film. This pattern of contrast corresponds to the internal structure of the object. Radiographs produced in this way are essentially still photographs of the internal structure of the Specimen. Areas of the film that are not exposed correspond to thick part of the Specimen or to parts of the Specimen that are made from elements with high atomic numbers, such as iron or lead.

Another type of X-radiography uses a fluorescent screen as the Target to produce visible light images of the internal structure of an object. The fluorescent screen is made of materials that emit visible light when X-rays strike the screen. The radiograph produced in visible light corresponds to a radiograph produced by exposure of film. Security systems at airports use fluorescent screens with television cameras to produce radiographs of baggage and other objects for quick inspection of their contents and/or internal structure. The Fluoroscope, a medical device, also uses a fluorescent screen to produce dynamic images that help doctors to diagnose problems that cannot be observed with the static images produced by film.

At the Goldwater Center at ASU we use a fluorescent screen and conventional polaroid film to produce images similar to traditional X-Radiographs (see Fig. 8.6). The fluorescent screen and polaroid film are assembled together such that the film to be exposed is pressed agaginst the screen and held in place by a cartridge that protects the polaroid film from exposure by ambient visible light. The cartridge, with the screen and film inside, is the Target. It is important to note that the polaroid film is sensitive to visible light only. X-Rays will not expose the polaroid film. When X-rays strike the fluorescent screen, it emits visible light in a pattern that conveys the pattern of X-rays on the screen. Hence, instead of exposing the film directly, the X-rays generate a visible light image on the fluorescent screen which then exposes the polaroid film in the traditional photographic way. The exposed polaroid film is then removed from the cartridge in a manner similar to a modern polaroid camera. The positive print is available in 1 minute.

Absorption of Light

The mechanism for the production of color by materials is the selective removal of certain wavelengths (or energies) of light from the electromagnetic spectrum of white light. The light must penetrate into the material and encounter energy-absorbing events for this selective removal to occur. A photon of given energy transfers all its energy to an electron in the absorbing material. The photon is "lost" from the light beam as it is absorbed, and the electron is excited to an available higher energy state by the gain in energy from the photon. The amount of absorption of photons depends on the energy levels available to the electron which

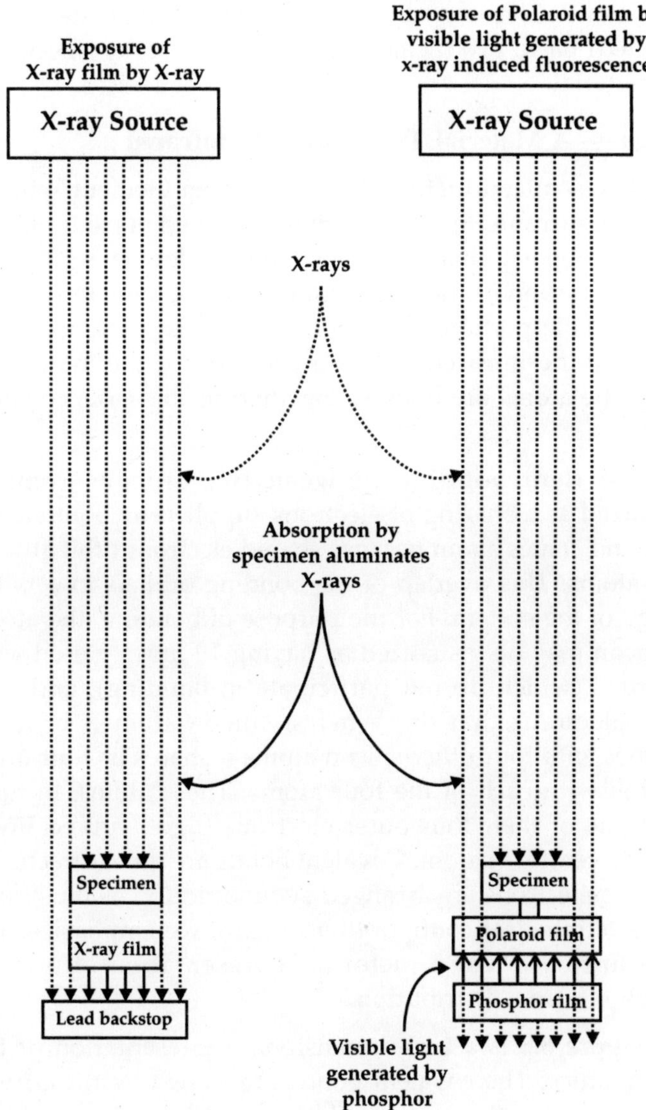
X-ray Radiography
Exposure of
X-ray film by X-ray
Exposure of Polaroid film by
visible light generated by
x-ray induced fluorescence
X-ray Source
X-ray Source
X-rays
Absorption by
specimen aluminates
X-rays
Specimen
X-ray film
Lead backstop
Specimen
Polaroid film
Phosphor film
Visible light
generated by
phosphor

Fig. 8.7

absorbs the photon energy. A photon is not absorbed if there are no energy levels available for photon-induced electron transitions. The non-absorbed photons are scattered (or reflected) back, producing the sensation of color specific to the material.

Silicon — A Material Transparent to Infrared

The semiconductor silicon (Si) is an element with 14 electrons surrounding the positively-charged nucleus. Silicon has the property that it is transparent to low energy in the infrared portion of the spectrum but it is opaque to photons in the visible portion of the spectrum. This transparency of silicon to the infrared (IR) is a property of the manner in which the atoms are bonded together; in this case, covalent bonds.

Covalent bonding between two silicon atoms is visualized as a sharing of electrons supplied by both atoms. The bond comes about because shared electrons orbit around both atoms. This overlap of the bonding orbitals lowers the energy of the system. For the purpose of bonding, the atoms of silicon may be visualized as having 10 inner, closed-shell electrons (which do not participate in bonding), and four outer electrons that do. When a silicon atom is brought together with four other silicon atoms it shares its four outer electron with each of the four atoms. These atoms, in turn, share one of their four outer electrons (light, curved lines) wiht the central Si atom. Covalent bonds are quite directional (the four electrons are arranged symmetrically) as illustrated by the tetrahedral configuration. Again, we emphasize that these lines are just a pictorial representation of a more diffuse electron distribution.

Figure 8.8 is a two-dimensional representation of the silicon lattice. The covalent bonds are shown by the curved lines connecting the Si atoms. Although this is a line-and-ball representation, it shows that all the outer or valence electrons are tied up in covalent bonds. The inner 10 electrons are tightly bound (binding energies greater than 100 eV) and the outer four form the covalent bonds. Consequently, there are essentially no free electrons available for charge transport.

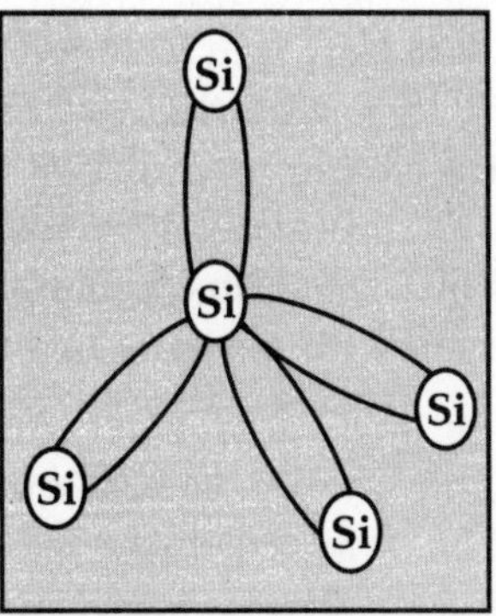

Fig. 8.8: Covalent bonds of silicon

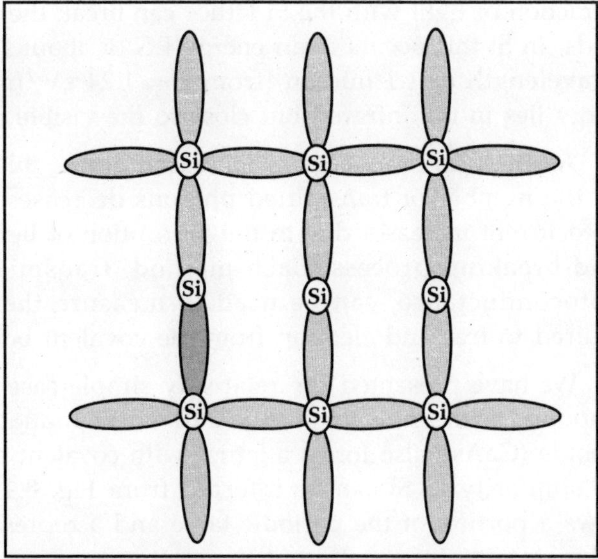

Fig. 8.9: Two-dimensional representation of the Si lattice for high-purity Si, where all the outer electrons are in covalent bonds.

If we shine light of a fixed energy on the sample, we can break the bonds if the photon energy is greater than the bond energy (about 1.1 eV for Si), which corresponds to the infrared portion of the spectrum. When a bond is broken, the liberated electron is now free to move within the crystal. The empty site, or hole, left by the escaping electron can be occupied by a nearby electron. Consequently, the hole can also migrate through the crystal by exchanges with the bound electrons. Both electrons and holes can trasnport charge leading to a current, a photocurrent.

If the sample is irradiated with low-energy photons which are not energetic enough to break the covalent bonds, the photons will be transmitted through the sample unattenuated (Fig. 8.9). There is no mechanism for the absorption of light as well as no photocurrent. As the energy of the photons is increased, a threshold is reached where the interaction of light with the Si lattice can break the covalent bonds. In Si this occurs at an energy EG of about 1.1 eV, at a wavelength of 1.1 micron [from E = 1.24 eV/(m)]. This energy lies in the infrared but close to the visible range.

As the photon energy is increased above this energy EG, the number of transmitted photons decreases and the photocurrent increases due to the absorption of light in the bond-breaking process. Each method, transmission or photoconductance, can be used to measure the energy required to free and electron from the covalent bonds.

We have presented the relatively simple case of Si. A compound semiconductor such as a III-V compound, gallium arsenide (GaAs), also forms a lattice with covalent bonding. The similarity to Si can be inferred from Fig. 8.11 which shows a portion of the periodic table and a representation of the atomic configuration of the electrons.

All five elements have 28 electrons in filled inner shells (K, L, andM shells) and electrons in the outer, less tightly bound shell (N shell). The column III element GA has three outer electrons, and the column V element As has five outer electrons. When these two elements are arranged in a lattice they share outer electrons to form covalent bonds with Ga atoms providing three electrons and As atoms, five electrons. There is, in effect, a Ga sublattice and an As sublattice.

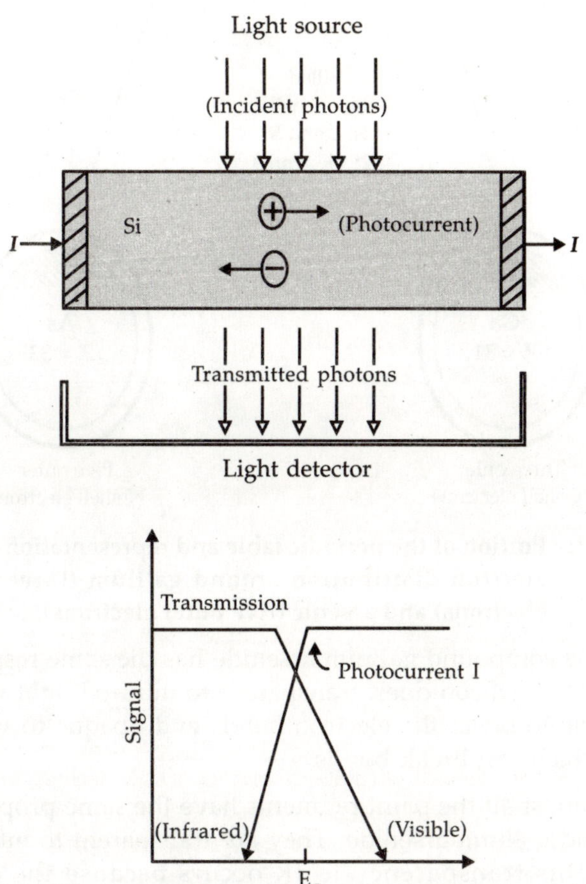

Fig. 8.10: Photons incident on a silicon sample, with the lower portion displaying the signal from the transmitted photons (not absorbed) and the photocurrent versus photon energy.

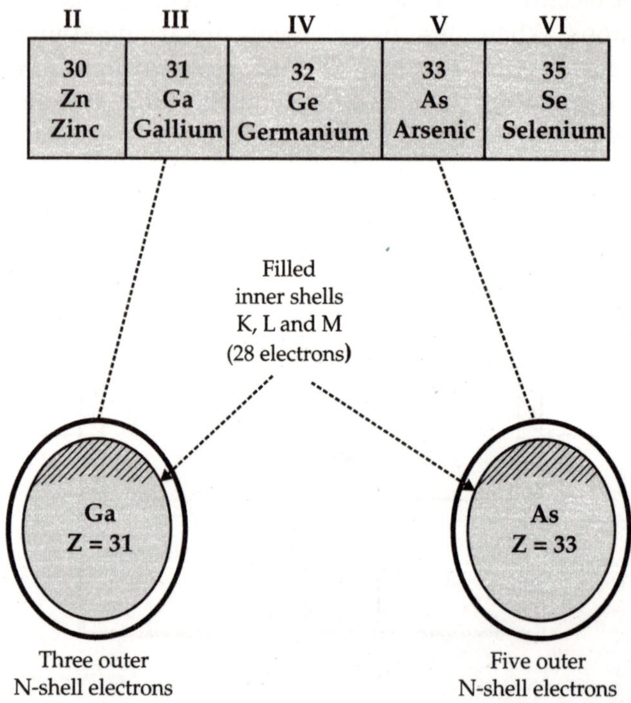

Fig. 8.11: Portion of the periodic table and representation of the electron distribution around gallium (three outer electrons) and arsenic (five outer electrons).

The compound gallium arsenide has the same response to light that silicon does: transparent to infrared light which is unable to break the electron bonds and opaque to visible light which can break bonds.

Almost all the paint pigments have the same properties as Si and gallium arsenide. They are transparent to infrared light. This transparency to IR occurs because the paint pigments are nearly all oxides (such as titanium white, titanium oxide) or sulfides (such as the red vermilion, mercury sulfide). In pure form, they are insulators or semiconductors with almost no electrons available for light absorption in the IR.

Pigment Response in the Infrared

In the infrared (IR) portion of the spectrum the wavelengths are greater than 700 nanometers with photon energies less than 1.8 electron-volts (eV). These infrared photon energies, typically around 1 eV, are so low that the photons are not absorbed by most pigments. The paint layers are therefore relatively transparent to infrared radiation. Infrared radiation penetrates the upper paint layers, but is absorbed by dark preliminary drawings made with charcoal that reside beneath them. The remaining radiation is reflected by white or light-colored grounds. With the aid of infrared sensitive photo equipment, the underdrawing can be seen because of the difference between absorbed and reflected radiation.

A particularly convincing demonstration of the transparency of pigments to the infrared is provided . In this case, an underdrawing is made using charcoal on a white ground. A layer of paint is then applied to the drawing which is viewed with an infrared sensitive video camera.

Fluorescence with Ultraviolet Light

Photons of visible light are not absorbed in layers of varnish and binding media in painting but are transmitted through them. The higher energy photons in the ultraviolet (UV) region of the spectrum are strongly absorbed in these varnishes and binding media.

The absorption of ultraviolet photons causes chemical reactions resulting, in some cases, in the emission of photns in the visible region of the spectrum. The absorption of light photns of high energy and re-emission of photons of lower energy, in the visible region, is commonly called fluorescence.

Figure 8.12: Illustration of the fluorescence process: An incident ultraviolet photon is absorbed and transfers its energy to an electron (process 1), the electron loses some of its energy in non-radiative processes, 2, and then the electron makes a transition to the ground state, process 3, emitting a photon with energy in the visible region.

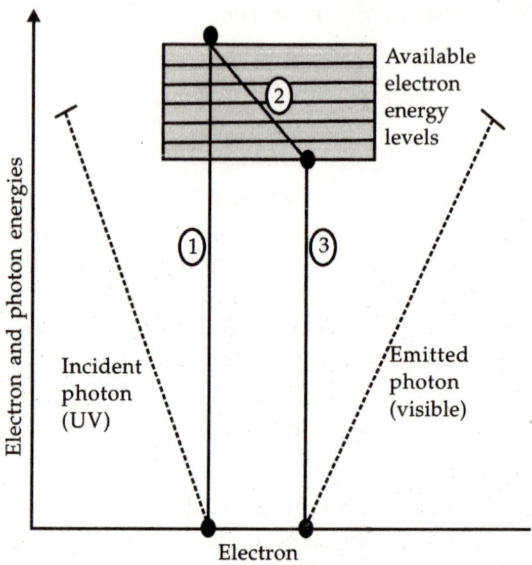

Fig. 8.12

The fluorescence process sketched in fig. 8.12 depicts an incident UV photon interacting with an electron. The photon is absorbed and transfers its energy to the electron raising it to one of the availabe energy levels. This enhanced energy of the electron may be partially lost by heat and then thte electron can make a transition back to the original state by emitting a photon The emitted photon is of a lower energy than the incident photon. Aged varnishes have different fluorescent properties than new varnishes and some modern pigments fluoresce differently than traditional pigments. Thus examination of a painting under ultraviolet light can often reveal retouched or restored areas.

CHAPTER 9

Photodissociation

Photodissociation, photolysis, or photodecomposition is a chemical reaction in which a chemical compound is broken down by photons. Photodissociation is not limited to visible light, but to have enough energy to break up a molecule, the photon is likely to be an electromagnetic wave with the energy of visible light or higher, such as ultraviolet light, x-rays and gamma rays. The direct process is defined as the interaction of one or more photons interacting with one target molecule.

Photolysis in Photosynthesis

Photolysis is a part of the light-dependent reactions of photosynthesis. The general reaction of photosynthetic photolysis can be given as:

H_2A + 2 photons (light) $\rightarrow 2e^- + 2H^+ + A$

The chemical nature of "A" depends on the type of organism. For example in purple sulfur bacteria, hydrogen sulfide (H_2S) is oxidized to sulfur (S). In oxygenic photosynthesis, water (H_2O) serves as a substrate for photolysis resulting in the generation of free oxygen (O_2).

This process is responsible for generating the majority of breathable oxygen in earth's atmosphere. Photolysis of water occurs in the thylakoids of cyanobacteria and the chloroplasts of green algae and plants.

Energy Transfer Models

The conventional, semi-classical, model describes the photosynthetic energy transfer process as one in which excitation energy hops from light-capturing pigment molecules to reaction center.

The effectiveness of photons of different wavelengths depends on the absorption spectra of the photosynthetic pigments in the organism. Chlorophylls absorb light in the violet-blue and red parts of the spectrum, while accessory pigments capture other wavelengths as well. The phycobilins of red algae absorb blue-green light which penetrates deeper into water than red light, enabling them to photosynthesize in deep waters. Each absorbed photon causes the formation of an exciton (an electron excited to a higher energy state) in the pigment molecule. The energy of the exciton is transferred to a chlorophyll molecule (P680, where P stands for pigment and 680 for its absorption maximum at 680 nm) in the reaction center of photosystem II via resonance energy transfer. P680 can also directly absorb a photon at a suitable wavelength.

Photolysis during photosynthesis occurs in a series of light-driven oxidation events. The energized electron (exciton) of P680 is captured by a primary electron acceptor of the photosynthetic electron transfer chain and thus exits photosystem II. In order to repeat the reaction, the electron in the reaction center needs to be replenished. This occurs by oxidation of water in the case of oxygenic photosynthesis. The electron-deficient reaction center of photosystem II (P680*) is the strongest biological oxidizing agent known on earth, which allows it to break apart molecules as stable as water.

The water-splitting reaction is catalyzed by the oxygen evolving complex of photosystem II. This protein-bound inorganic complex contains four manganese ions, plus a calcium and chloride ion as cofactors. Two water molecules are complexed by the manganese cluster, which then undergoes a series of four electron removals (oxidations) to replenish the reaction center of photosystem II. At the end of this cycle, free oxygen (O_2) is generated and the hydrogen of the water molecules has been converted to four protons released into the thylakoid lumen.

These protons, as well as additional protons pumped across the thylakoid membrane coupled with the electron transfer chain, form a proton gradient across the membrane that drives photophosphorylation and thus the generation of chemical energy in the form of adenosine triphosphate (ATP). The electrons reach the P700 reaction center of photosystem I where they are energized again by light. They are passed down another electron transfer chain and finally combine with the coenzyme NADP+ and protons outside the thylakoids to NADPH. Thus, the net oxidation reaction of water photolysis can be written as:

$$2H_2O + 2NADP^+ + 8 \text{ photons (light)} \rightarrow 2NADPH + 2H^+ + O_2$$

The free energy change (ΔG) for this reaction is 102 kilocalories per mole. Since the energy of light at 700 nm is about 40 kilocalories per mole of photons, approximately 320 kilocalories of light energy are available for the reaction. Therefore, approximately one-third of the available light energy is captured as NADPH during photolysis and electron transfer. An equal amount of ATP is generated by the resulting proton gradient. Oxygen as a byproduct is of no further use to the reaction and thus released into the atmosphere.

Quantum Models

In 2007 a quantum model was proposed by Graham Fleming, which includes the possibility that photosynthetic energy transfer might involve quantum oscillations, explaining its unusually high efficiency.

According to Fleming there is direct evidence that remarkably long-lived wavelike electronic quantum coherence plays an important part in energy transfer processes during photosynthesis, which can explain the extreme efficiency of the energy transfer because it enables the system to sample all the potential energy pathways, with low loss, and choose the most efficient one.

Photolysis in the Atmosphere

Photolysis also occurs in the atmosphere as part of a series of reactions by which primary pollutants such as hydrocarbons and nitrogen oxides react to form secondary pollutants such as peroxyacyl nitrates.

The two most important photodissociaton reactions in the troposphere are firstly:

$$O_3 + h\lambda \rightarrow O_2 + O(^1D) \rightarrow \lambda\ 320\ nm$$

which generates an excited oxygen atom which can go on to react with water to give the hydroxyl radical:

$$O(^1D) + H_2O \rightarrow 2OH$$

The hydroxyl radical is central to atmospheric chemistry as it initiates the oxidation of hydrocarbons in the atmosphere and so acts like a detergent.

Secondly the reaction:

$$NO_2 + h\lambda \rightarrow NO + O$$

is a key reaction in the formation of tropospheric ozone.

The formation of the ozone layer is also caused by photodissociation. Ozone in the earth's stratosphere is created by ultraviolet light striking oxygen molecules containing two oxygen atoms (O_2), splitting them into individual oxygen atoms (atomic oxygen); the atomic oxygen then combines with unbroken O_2 to create ozone, O_3. In addition, photolysis is the process by which CFCs are broken down in the upper atmosphere to form ozone-destroying chlorine free radicals.

Astrophysics

In astrophysics, photodissociation is one of the major processes through which molecules are broken down (but new molecules are being formed). Because of the vacuum of the interstellar medium, molecules and free radicals can exist for a long time. Photodissociation is the main path by which molecules are broken down. Photodissociation rates are very important in the study of the composition of interstellar clouds in which stars are formed.

Typical examples of photodissociation in the interstellar medium are (hυ is the scientific notation for light, specifically a photon):

$$H_2O + hv \rightarrow H + OH$$

$$CH_4 + hv \rightarrow CH_3 + H$$

Multiple Photon Dissociation

In comparison to ultraviolet or other high energy photons, single photons in the infrared spectral range usually are not energetic enough for direct photodissociation of molecules. However, after absorption of multiple infrared photons a molecule may gain internal energy to overcome its barrier for dissociation. Multiple photon dissociation (MPD) can be achieved by applying high power lasers, e.g. a Carbon dioxide laser, or a Free electron laser, or by long interaction times of the molecule with the radiation field without the possibility for rapid cooling, e.g. by collisions. The latter method allows even for MPD induced by black body radiation.

LIGHT-INDEPENDENT REACTION

Photosynthesis, the light-independent reactions, are chemical reactions that convert carbon dioxide and other compounds into glucose. They occur in the stroma, the fluid filled area of a chloroplast outside of the thylakoid membranes. These reactions take the products of the light-dependent reactions and perform further chemical processes on them. There are three phases to the light-independent reactions, collectively called the Calvin Cycle: Carbon Fixation, Reduction reactions, and ribulose 1,5-diphosphate (RuDP) regeneration.

The light-independent reactions are sometimes referred to as the dark reactions, though that term may be misleading as they do not actually require darkness to proceed. The term "light-independent" is used to emphasize that the reactions occur regardless of the amount of light present as long as the proper substrate compounds are available. Even this term can be criticized, however, as the availability of substrates in plants depends on photosynthesis, so the reactions cannot be said to be entirely "light-independent." Furthermore, the term "dark reactions" may be more accurate in CAM (Crassulacean acid metabolism) plants, which only take up CO_2, which is necessary for the reactions, to proceed at night.

The carbon fixation reaction is the first step of the light-independent reactions. Carbon from carbon dioxide is "fixed" into a larger carbohydrate via one of three three chemical pathways: C3 carbon fixation (the most common), C4 carbon fixation, and CAM (Crassulacean Acid Metabolism). C3 fixation occurs as the first step of the Calvin-Benson cycle in all plants. C4 plants first fix carbon dioxide into oxaloacetate, a 4C acid, and then into malate, which is transported to the bundle sheath cells, where it is decarboxylated to supply carbon dioxide to the Calvin-Benson cycle. CAM plants perform a similar process, but instead of spatial separation of carbon dioxide fixation and the Calvin-Benson cycle, they separate the processes in time: at night they open their stomata and fix carbon dioxide into oxaloacetate which is converted to malate. During the day malate is decarboxylated and the thus released carbon dioxide enters the Calvin-Benson cycle.

PHOTOSYNTHESIS

Photosynthesis is a metabolic pathway that converts carbon dioxide into organic compounds, especially sugars, using the energy from sunlight. Photosynthesis occurs in plants, algae, and many species of Bacteria, but not in Archaea. Photosynthetic organisms are called photoautotrophs, but not all organisms that use light as a source of energy carry out photosynthesis, since photoheterotrophs use organic compounds, rather than carbon dioxide, as a source of carbon In plants, algae and cyanobacteria photosynthesis uses carbon dioxide and water, releasing oxygen as a waste product. Photosynthesis is crucially important for life on Earth, since as well as it maintaining the normal level of oxygen in the atmosphere, nearly all life either depends on it directly as a source of energy or indirectly as the ultimate source of the energy in their food The amount of energy trapped by photosynthesis is immense, approximately 100 terawatts per year which is about seven times larger than the yearly power consumption of human civilization In all, photosynthetic organisms convert around 10,000,000,000 tonnes of carbon into biomass per year.

Although photosynthesis can occur in different ways in different species, some features are always the same. For example, the process always begins when energy from light is absorbed by proteins called photosynthetic reaction centers that contain chlorophyll. In plants these proteins are held inside organelles called chloroplasts, while in bacteria they are embedded in the plasma membrane. Some of the light energy gathered by chlorophylls is stored in the form of adenosine triphosphate (ATP). The rest of the energy is used to remove electrons from a substance such as water. These electrons are then used in the reactions that turn carbon dioxide into organic compounds. In plants, algae and cyanobacteria these reactions are called the Calvin cycle, but different sets of reactions can be found in bacteria, such as the reverse Krebs cycle in Chlorobium. Many photosynthetic organisms have adaptations that concentrate or store carbon dioxide, this helps reduce a wasteful process called photorespiration that would otherwise consume part of the sugar produced during photosynthesis.

Photosynthesis evolved early in the evolutionary history of life, when all forms of life on Earth were microorganisms. Although the dates are difficult to estimate with any accuracy, the first photosynthetic organisms probably evolved about 3,500 million years ago, and used hydrogen or hydrogen sulfide as sources of electrons, rather than water. Cyanobacteria appeared later, around 3,000 million years ago, and changed the Earth forever when they began to oxygenate the atmosphere, beginning about 2,400 million years ago This new atmosphere allowed the evolution of complex life such as protists. Eventually, about 550 million years ago, one of these protists formed a symbiotic relationship with a cyanobacterium, producing the ancestor of the plants and algae The chloroplasts in modern plants are the descendants of these ancient symbiotic cyanobacteria.

Overview

Photosynthetic organisms are photoautotrophs, which means that they are able to synthesize food directly from carbon dioxide using energy from light. In plants algae and cyanobacteria, photosynthesis releases oxygen, this is called oxygenic photosynthesis. Although there are somedifferences between oxygenic photosynthesis in plants, algae and cyanobacteria, the overall process is quite similar in these organisms. However, there are some types of bacteria that carry out anoxygenic photosynthesis, which consumes carbon dioxide but does not release oxygen.

Carbon dioxide is converted into sugars in a process called carbon fixation. Carbon fixation is a redox reaction, so photosynthesis needs to supply both a source of energy to drive this process, and also the electrons needed to convert carbon dioxide into carbohydrate which is a reduction reaction. In general outline, photosynthesis is the opposite of cellular respiration, where glucose and other compounds are oxidized to produce carbon dioxide, water, and release chemical energy. However, the two processes take place through a different sequence of chemical reactions and in different cellular compartments.

The general equation for photosynthesis is therefore:

$CO_2 + 2\ H_2A + \text{photons} \rightarrow (CH_2O)_n + H_2O + 2A$

carbon dioxide + electron donor + light energy → carbohydrate + oxygen + oxidized electron donor

Since water is used as the electron donor in oxygenic photosynthesis, the equation for this process is:

$CO_2 + 2\ H_2O + \text{photons} \rightarrow (CH_2O)_n + H_2O + O_2$

carbon dioxide + water + light energy →carbohydrate + oxygen + water

Photosynthesis occurs in two stages. In the first stage, light-dependent reactions or light reactions capture the energy of light and use it to make the energy-storage molecules ATP and NADPH. During the second stage, the light-independent reactions use these products to capture and reduce carbon dioxide.

Photosynthetic Membranes and Organelles

The proteins that gather light for photosynthesis are embedded within cell membranes. The simplest way these are arranged is in photosynthetic bacteria, where these proteins are held within the plasma membrane However, this membrane may be tightly-folded into cylindrical sheets called thylakoids or bunched up into round vesicles called *intracytoplasmic membranes*. These structures can fill most of the interior of a cell, giving the membrane a very large surface area and therefore increasing the amount of light that the bacteria can absorb.

In plants and algae, photosynthesis takes place in organelles called chloroplasts. A chloroplast is contained by an envelope that consists of an inner and an outer phospholipid membrane. Between these two layers is the intermembrane space. A typical plant cell contains about 10 to 100 chloroplasts. Within the stroma are stacks of thylakoids, the sub-organelles which are the site of photosynthesis. The thylakoids are arranged in stacks called grana (singular: granum). A thylakoid has a flattened disk shape. Inside it is an empty area called the thylakoid space or lumen. The thylakoid membrane contains many integral and peripheral membrane proteins. The proteins complexes which contain special pigments absorbing light energy are called photosystems.

Plants absorb light primarily using the pigment chlorophyll, which is the reason that most plants have a green color. Besides chlorophyll plants also use pigments such as *carotenes* and *xanthophylls*. Algae also use chlorophyll, but various other pigments are present as phycocyanin, carotenes, and xanthophylls in green algae, phycoerythrin in red algae (rhodophytes) and fucoxanthol in brown algae and diatoms resulting in a wide variety of colors.

These pigments are embedded in plants and algae in special antenna-proteins. In such proteins all the pigments are ordered to work well together. Such a protein is also called a *light-harvesting complex*.

Although all cells in the green parts of a plant have chloroplasts, most of the energy is captured in the leaves. The cells in the interior tissues of a leaf, called the *mesophyll*, can contain between 450,000 and 800,000 chloroplasts for every square millimetre of leaf. The surface of the leaf is uniformly coated with a water-resistant waxy cuticle that protects the leaf from excessive evaporation of water and decreases the absorption of ultraviolet or blue light to reduce heating. The transparent epidermis layer allows light to pass through to the palisade mesophyll cells where most of the photosynthesis takes place.

Light Reactions

In the light reactions, one molecule of the pigment chlorophyll absorbs one photon and loses one electron. This electron is passed to a modified form of chlorophyll called *pheophytin*, which passes the electron to a quinone molecule, allowing the start of a flow of electrons down an electron transport chain that leads to the ultimate reduction of NADP to NADPH. In addition, this creates a proton gradient across the chloroplast membrane; its dissipation is used by ATP Synthase for the concomitant synthesis of ATP. The chlorophyll molecule regains the lost electron from a water molecule through a process called photolysis, which releases a dioxygen (O_2) molecule. The overall equation for the light-dependent reactions under the conditions of non-cyclic electron flow in green plants is:

$$2\ H_2O + 2\ NADP^+ + 2\ ADP + 2\ P_i + \text{light} \rightarrow 2\ NADPH + 2\ H^+ + 2\ ATP + O_2$$

Not all wavelengths of light can support photosynthesis. The photosynthetic action spectrum depends on the type of accessory pigments present. For example, in green plants, the action spectrum resembles the absorption spectrum for chlorophylls and carotenoids with peaks for violet-blue and red light. In red algae, the action spectrum overlaps with the absorption spectrum of phycobilins for blue-green light, which allows these algae to grow in deeper waters that filter out the longer wavelengths used by green plants. The non-absorbed part of the light spectrum is what gives photosynthetic organisms their color (e.g., green plants, red algae, purple bacteria) and is the least effective for photosynthesis in the respective organisms.

Z Scheme

In plants, light-dependent reactions occur in the thylakoid membranes of the chloroplasts and use light energy to synthesize ATP and NADPH. The light-dependent reaction has two forms; cyclic and non-cyclic reaction. In the non-cyclic reaction, the photons are captured in the light-harvesting antenna complexes of photosystem II by chlorophyll and other accessory pigments (see diagram at right). When a chlorophyll molecule at the core of the photosystem II reaction center obtains sufficient excitation energy from the adjacent antenna pigments, an electron is transferred to the primary electron-acceptor molecule, Pheophytin, through a process called Photoinduced charge separation. These electrons are shuttled through an electron transport chain, the so called Z-scheme shown in the diagram, that initially functions to generate a chemiosmotic potential across the membrane. An ATP synthase enzyme uses the chemiosmotic potential to make ATP during photophosphorylation, whereas NADPH is a product of the terminal redox reaction in the Z-scheme. The electron enters the Photosystem I molecule. The electron is excited due to the light absorbed by the photosystem. A second electron carrier accepts the electron, which again is passed down lowering energies of electron acceptors. The energy created by the electron acceptors is used to move hydrogen ions across the thylakoid membrane into the lumen. The electron is used to reduce the co-enzyme NADP, which has functions in the light-independent reaction. The cyclic reaction is similar to that of the non-cyclic, but differs in the form that it generates only ATP, and no reduced NADP (NADPH) is created. The cyclic reaction takes place only at photosystem I. Once the electron is displaced from the photosystem, the electron is passed down the electron acceptor molecules and returns back to photosystem I, from where it was emitted, hence the name cyclic reaction.

Water Photolysis

The NADPH is the main reducing agent in chloroplasts, providing a source of energetic electrons to other reactions. Its production leaves chlorophyll with a deficit of electrons (oxidized), which must be obtained from some other reducing agent. The excited electrons lost from chlorophyll in photosystem I are replaced from the electron transport chain by plastocyanin. However, since photosystem II includes the first steps of the Z-scheme, an external source of electrons is required to reduce its oxidized chlorophyll a molecules. The source of electrons in green-plant and cyanobacterial photosynthesis is water. Two water molecules are oxidized by four successive charge-separation reactions by photosystem II to yield a molecule of diatomic oxygen and four hydrogen ions; the electron yielded in each step is transferred to a redox-active tyrosine residue that then reduces the photoxidized paired-chlorophyll a species called P680 that serves as the primary (light-driven) electron donor in the photosystem II reaction center. The oxidation of water is catalyzed in photosystem II by a redox-active structure that contains four manganese ions and a calcium ion; this oxygen-evolving complex binds two water molecules and stores the four oxidizing equivalents that are required to drive the water-oxidizing reaction. Photosystem II is the only known biological enzyme that carries out this oxidation of water. The hydrogen ions contribute to the transmembrane chemiosmotic potential that leads to ATP synthesis. Oxygen is a waste product of light-independent reactions, but the majority of organisms on Earth use oxygen for cellular respiration, including photosynthetic organisms.

OXYGEN AND PHOTOSYNTHESIS

With respect to oxygen and photosynthesis, there are two important concepts.

1. Plant and cyanobacterial (blue-green algae) cells also use oxygen for cellular respiration, although they have a net output of oxygen since much more is produced during photosynthesis.

2. Oxygen is a product of the light-driven water-oxidation reaction catalyzed by photosystem II; it is not generated by the fixation of carbon dioxide. Consequently, the source of oxygen during photosynthesis is water, not carbon dioxide.

The concept that oxygen production is not directly associated with the fixation of carbon dioxide was first proposed by Cornelis Van Niel in the 1930s, who studied photosynthetic bacteria. Aside from the cyanobacteria, bacteria only have one photosystem and use reducing agents other than water. They get electrons from a variety of different inorganic chemicals including sulfide or hydrogen, so for most of these bacteria oxygen is not produced.

Others, such as the halophiles (an Archaea), produced so-called purple membranes where the bacteriorhodopsin could harvest light and produce energy. The purple membranes was one of the first to be used to demonstrate the chemiosmotic theory: light hit the membranes and the pH of the solution that contained the purple membranes dropped as protons were pumping out of the membrane.

Light-independent Reactions

The Calvin Cycle

In the Light-independent or dark reactions the enzyme RuBisCO captures CO_2 from the atmosphere and in a process that requires the newly formed NADPH, called the Calvin-Benson Cycle, releases three-carbon sugars, which are later combined to form sucrose and starch. The overall equation for the light-independent reactions in green plants is:

$$3\,CO_2 + 9\,ATP + 6\,NADPH + 6\,H^+ \rightarrow C_3H_6O_3\text{-phosphate} + 9\,ADP + 8\,Pi + 6\,NADP^+ + 3\,H_2O$$

To be more specific, carbon fixation produces an intermediate product, which is then converted to the final carbohydrate products. The carbon skeletons produced by photosynthesis are then variously used to form other organic compounds, such as the building material cellulose, as precursors for lipid and amino acid biosynthesis, or as a fuel in cellular respiration. The latter occurs not only in plants but also in animals when the energy from plants gets passed through a food chain.

The fixation or reduction of carbon dioxide is a process in which carbon dioxide combines with a five-carbon sugar, ribulose 1, 5-bisphosphate (RuBP), to yield two molecules of a three-carbon compound, glycerate 3-phosphate (GP), also known as 3-phosphoglycerate (PGA). GP, in the presence of ATP and NADPH from the light-dependent stages, is reduced to glyceraldehyde 3-phosphate (G3P). This product is also referred to as 3-phosphoglyceraldehyde (PGAL) or even as triose phosphate. Triose is a 3-carbon sugar (see carbohydrates). Most (5 out of 6 molecules) of the G3P produced is used to regenerate RuBP so the process can continue. The 1 out of 6 molecules of the triose phosphates not "recycled" often condense to form hexose phosphates, which ultimately yield sucrose, starch and cellulose. The sugars produced during carbon metabolism yield carbon skeletons that can be used for other metabolic reactions like the production of amino acids and lipids.

C_4 AND C_3 PHOTOSYNTHESIS AND CAM

In hot and dry conditions, plants will close their stomata to prevent loss of water. Under these conditions, CO_2 will decrease, and dioxygen gas, produced by the light reactions of photosynthesis, will increase in the leaves, causing an increase of photorespiration by the oxygenase activity of ribulose-1, 5-bisphosphate carboxylase/oxygenase and decrease in carbon fixation. Some plants have evolved mechanisms to increase the CO_2 concentration in the leaves under these conditions.

C_4 Carbon Fixation

C_4 plants chemically fix carbon dioxide in the cells of the mesophyll by adding it to the three-carbon molecule phosphoenolpyruvate (PEP), a reaction catalyzed by an enzyme called PEP carboxylase and which creates the four-carbon organic acid, oxaloacetic acid. Oxaloacetic acid or malate synthesized by this process is then translocated to specialized bundle sheath cells where the enzyme, rubisco, and other Calvin cyle enzymes are located, and where CO_2

released by decarboxylation of the four-carbon acids is then fixed by rubisco activity to the three-carbon sugar 3 phosphoglyceric acids. The physical separation of rubisco from the oxygen-generating light reactions reduces photorespiration and increases CO_2 fixation and thus photosynthetic capacity of the leaf. C_4 plants can produce more sugar than C_3 plants in conditions of high light and temperature. Many important crop plants are C_4 plants including maize, sorghum, sugarcane, and millet. Plants lacking PEP-carboxylase are called C_3 plants because the primary carboxylation reaction, catalyzed by Rubiso, produces the three-carbon sugar 3 phosphoglyceric acids directly in the Calvin-Benson Cycle.

Xerophytes such as cacti and most succulents also use PEP carboxylase to capture carbon dioxide in a process called Crassulacean acid metabolism (CAM). In contrast to C4 metabolism, which physically separates the CO_2 fixation to PEP from the Calvin cycle, CAM only temporally separates these two processes. CAM plants have a different leaf anatomy than C4 plants, and fix the CO_2 at night, when their stomata are open. CAM plants store the CO_2 mostly in the form of malic acid via carboxylation of phosphoenolpyruvate to oxaloacetate, which is then reduced to malate. Decarboxylation of malate during the day releases CO_2 inside the leaves thus allowing carbon fixation to 3-phosphoglycerate by rubisco.

Order and Kinetics

The overall process of photosynthesis takes place in four stages. The first, energy transfer in antenna chlorophyll takes place in the femtosecond [1 femtosecond (fs) = 10^{-15} s] to picosecond [1 picosecond (ps) = 10^{-12} s] time scale. The next phase, the transfer of electrons in photochemical reactions, takes place in the picosecond to nanosecond time scale [1 nanosecond (ns) = 10^{-9} s]. The third phase, the electron transport chain and ATP synthesis, takes place on the microsecond [1 microsecond (μs) = 10^{-6} s] to millisecond [1 millisecond (ms) = 10^{-3} s) time scale. The final phase is carbon fixation and export of stable products and takes place in the millisecond to second time scale. The first three stages occur in the thylakoid membranes.

Efficiency

Plants convert light into chemical energy with a maximum photosynthetic efficiency of approximately 6% By comparison solar panels convert light into electric energy at a photosynthetic efficiency of approximately 10-20%. Actual plant's photosynthetic efficiency varies with the frequency of the light being converted, light intensity, temperature and proportion of CO_2 in atmosphere.

Evolution

Early photosynthetic systems, such as those from green and purple sulfur and green and purple non-sulfur bacteria, are thought to have been anoxygenic, using various molecules as electron donors. Green and purple sulfur bacteria are thought to have used hydrogen and sulfur as an electron donor. Green nonsulfur bacteria used various amino and other organic acids. Purple nonsulfur bacteria used a variety of non-specific organic molecules. The use of these molecules is consistent with the geological evidence that the atmosphere was highly reduced at that time.

The main source of oxygen in the atmosphere is oxygenic photosynthesis, and its first appearance is sometimes referred to as the oxygen catastrophe. Geological evidence suggests that oxygenic photosynthesis, such as that in cyanobacteria, became important during the Paleoproterozoic era around 2 billion years ago. Modern photosynthesis in plants and most photosynthetic prokaryotes is oxygenic. Oxygenic photosynthesis uses water as an electron donor which is oxidized to molecular dioxygen (O_2) in the photosynthetic reaction center.

Symbiosis and the Origin of Chloroplasts

Several groups of animals have formed symbiotic relationships with photosynthetic algae. These are most common in corals, sponges and sea anemones, possibly due to these animals having particularly simple body plans and large surface areas compared to their volumes In addition, a few marine molluscs Elysia viridis and Elysia chlorotica also maintain a symbiotic relationship with chloroplasts that they capture from the algae in their diet and then store in their bodies. This allows the molluscs to survive solely by photosynthesis for several months at a time Some of the genes from the plant cell nucleus have even been transferred to the slugs, so that the chloroplasts can be supplied with proteins that they need to survive.

An even closer form of symbiosis may explain the origin of chloroplasts. Chloroplasts have many similarities with photosynthetic bacteria including a circular chromosome, prokaryotic-type ribosomes, and similar proteins in the photosynthetic reaction cente The endosymbiotic theory suggests that photosynthetic bacteria were acquired (by endocytosis) by early eukaryotic cells to form the first plant cells. Therefore, chloroplasts may be photosynthetic bacteria that adapted to life inside plant cells. Like mitochondria, chloroplasts still possess their own DNA, separate from the nuclear DNA of their plant host cells and the genes in this chloroplast DNA resemble those in cyanobacteria DNA in chloroplasts codes for redox proteins such as photosynthetic reaction centers. The CoRR Hypothesis proposes that this Co-location is required for Redox Regulation.

Cyanobacteria and the Evolution of Photosynthesis

The biochemical capacity to use water as the source for electrons in photosynthesis evolved once, in a common ancestor of extant cyanobacteria. The geological record indicates that this transforming event took place early in Earth's history, at least 2450-2320 million years ago (Ma), and possibly much earlier. Available evidence from geobiological studies of Archean (>2500 Ma) sedimentary rocks indicates that life existed 3500 Ma, but the question of when oxygenic photosynthesis evolved is still unanswered. A clear paleontological window on cyanobacterial evolution opened about 2000 Ma, revealing an already-diverse biota of blue-greens. Cyanobacteria remained principal primary producers throughout the Proterozoic Eon (2500-543 Ma), in part because the redox structure of the oceans favored photoautotrophs capable of nitrogen fixation.[citation needed] Green algae joined blue-greens as major primary producers on continental shelves near the end of the Proterozoic, but only with the Mesozoic (251-65 Ma) radiations of dinoflagellates, coccolithophorids, and diatoms did primary production in marine shelf waters take modern form. Cyanobacteria remain critical to marine ecosystems as primary producers in oceanic gyres, as agents of biological nitrogen fixation, and, in modified form, as the plastids of marine algae.

Although some of the steps in photosynthesis are still not completely understood, the overall photosynthetic equation has been known since the 1800s.

Jan van Helmont began the research of the process in the mid-1600s when he carefully measured the mass of the soil used by a plant and the mass of the plant as it grew. After noticing that the soil mass changed very little, he hypothesized that the mass of the growing plant must come from the water, the only substance he added to the potted plant. His hypothesis was partially accurate—much of the gained mass also comes from carbon dioxide as well as water. However, this was a signaling point to the idea that the bulk of a plant's biomass comes from the inputs of photosynthesis, not the soil itself.

Joseph Priestley, a chemist and minister, discovered that when he isolated a volume of air under an inverted jar, and burned a candle in it, the candle would burn out very quickly, much before it ran out of wax. He further discovered that a mouse could similarly "injure" air. He then showed that the air that had been "injured" by the candle and the mouse could be restored by a plant.

In 1778, Jan Ingenhousz, court physician to the Austrian Empress, repeated Priestley's experiments. He discovered that it was the influence of sunlight on the plant that could cause it to rescue a mouse in a matter of hours.

In 1796, Jean Senebier, a Swiss pastor, botanist, and naturalist, demonstrated that green plants consume carbon dioxide and release oxygen under the influence of light. Soon afterwards, Nicolas-Théodore de Saussure showed that the increase in mass of the plant as it grows could not be due only to uptake of CO_2, but also to the incorporation of water. Thus the basic reaction by which photosynthesis is used to produce food (such as glucose) was outlined.

Cornelis Van Niel made key discoveries explaining the chemistry of photosynthesis. By studying purple sulfur bacteria and green bacteria he was the first scientist to demonstrate that photosynthesis is a light-dependent redox reaction, in which hydrogen reduces carbon dioxide.

Robert Emerson discovered two light reactions by testing plant productivity using different wavelengths of light. With the red alone, the light reactions were suppressed. When blue and red were combined, the output was much more substantial. Thus, there were two photosystems, one aborbing up to 600 nm wavelengths, the other up to 700. The former is known as PSII, the latter is PSI. PSI contains only chlorophyll a, PSII contains primarily chlorophyll a with most of the available chlorophyll b, among other pigments.

Further experiments to prove that the oxygen developed during the photosynthesis of green plants came from water, were performed by Robert Hill in 1937 and 1939. He showed that isolated chloroplasts give off oxygen in the presence of unnatural reducing agents like iron oxalate, ferricyanide or benzoquinone after exposure to light. The Hill reaction is as follows:

$$2\ H_2O + 2\ A + (\text{light, chloroplasts}) \rightarrow 2\ AH_2 + O_2$$

where A is the electron acceptor. Therefore, in light the electron acceptor is reduced and oxygen is evolved. Cyt b6, now known as a plastoquinone, is one electron acceptor.

Samuel Ruben and Martin Kamen used radioactive isotopes to determine that the oxygen liberated in photosynthesis came from the water.

Melvin Calvin and Andrew Benson, along with James Bassham, elucidated the path of carbon assimilation (the photosynthetic carbon reduction cycle) in plants. The carbon reduction cycle is known as the Calvin cycle, which inappropriately ignores the contribution of Bassham and Benson. Many scientists refer to the cycle as the Calvin-Benson Cycle, Benson-Calvin, and some even call it the Calvin-Benson-Bassham (or CBB) Cycle.

A Nobel Prize winning scientist, Rudolph A. Marcus, was able to discover the function and significance of the electron transport chain.

Factors

There are three main factors affecting photosynthesis and several corollary factors. The three main are:

1. Light irradiance and wavelength.
2. Carbon dioxide concentration.
3. Temperature.

Light Intensity (irradiance), Wavelength and Temperature

In the early 1900s Frederick Frost Blackman along with Gabrielle Matthaei investigated the effects of light intensity (irradiance) and temperature on the rate of carbon assimilation. At constant temperature, the rate of carbon assimilation varies with irradiance, initially increasing as the irradiance increases. However at higher irradiance this relationship no longer holds and the rate of carbon assimilation reaches a plateau.

At constant irradiance, the rate of carbon assimilation increases as the temperature is increased over a limited range. This effect is only seen at high irradiance levels. At low irradiance, increasing the temperature has little influence on the rate of carbon assimilation.

These two experiments illustrate vital points: firstly, from research it is known that photochemical reactions are not generally affected by temperature. However, these experiments clearly show that temperature affects the rate of carbon assimilation, so there must be two sets of reactions in the full process of carbon assimilation. These are of course the light-dependent 'photochemical' stage and the light-independent, temperature-dependent stage. Second, Blackman's experiments illustrate the concept of limiting factors. Another limiting factor is the wavelength of light. Cyanobacteria, which reside several meters underwater, cannot receive the correct wavelengths required to cause photoinduced charge separation in conventional photosynthetic pigments. To combat this problem, a series of proteins with different pigments surround the reaction center.This unit is called a phycobilisome.

Carbon Dioxide Levels and Photorespiration

As carbon dioxide concentrations rise, the rate at which sugars are made by the light-independent reactions increases until limited by other factors. RuBisCO, the enzyme that captures carbon dioxide in the light-independent reactions, has a binding affinity for both carbon dioxide and oxygen. When the concentration of carbon dioxide is high, RuBisCO will fix carbon dioxide. However, if the oxygen concentration is high, RuBisCO will bind oxygen instead of carbon dioxide. This process, called photorespiration, uses energy, but does not make sugar.

RuBisCO oxygenase activity is disadvantageous to plants for several reasons:

1. One product of oxygenase activity is phosphoglycolate (2 carbon) instead of 3-phosphoglycerate (3 carbon). Phosphoglycolate cannot be metabolized by the Calvin-Benson cycle and represents carbon lost from the cycle. A high oxygenase activity, therefore, drains the sugars that are required to recycle ribulose 5-bisphosphate and for the continuation of the Calvin-Benson cycle.
2. Phosphoglycolate is quickly metabolized to glycolate that is toxic to a plant at a high concentration; it inhibits photosynthesis.
3. Salvaging glycolate is an energetically expensive process that uses the glycolate pathway and only 75% of the carbon is returned to the Calvin-Benson cycle as 3-phosphoglycerate.

   A highly-simplified summary is:

   2 glycolate + ATP $\rightarrow$3-phosphoglycerate + carbon dioxide + ADP +NH_3

The salvaging pathway for the products of RuBisCO oxygenase activity is more commonly known as photorespiration, since it is characterized by light-dependent oxygen consumption and the release of carbon dioxide.

CHAPTER 10

Single Photon Emission Computed Tomography

Single photon emission computed tomography (SPECT, or less commonly, SPET) is a nuclear medicine tomographic imaging technique using gamma rays. It is very similar to conventional nuclear medicine planar imaging using a gamma camera. However, it is able to provide true 3D information. This information is typically presented as cross-sectional slices through the patient, but can be freely reformatted or manipulated as required.

Principles

In the same way that a plain X-ray is a 2-dimensional (2-D) view of a 3-dimensional structure, the image obtained by a gamma camera is a 2-D view of 3-D distribution of a radionuclide.

SPECT imaging is performed by using a gamma camera to acquire multiple 2-D images (also called projections), from multiple angles. A computer is then used to apply a tomographic reconstruction algorithm to the multiple projections, yielding a 3-D dataset. This dataset may then be manipulated to show thin slices along any chosen axis of the body, similar to those obtained from other tomographic techniques, such as MRI, CT, and PET.

SPECT is similar to PET in its use of radioactive tracer material and detection of gamma rays. In contrast with PET, however, the tracer used in SPECT emits gamma radiation that is measured directly, whereas PET tracer emits positrons which annihilate with electrons up to a few millimeters away, causing two gamma photons to be emitted in opposite directions. A PET scanner detects these emissions "coincident" in time, which provides more radiation event localization information and thus higher resolution images than SPECT (which has about 1 cm). SPECT scans, however, are significantly less expensive.

Because SPECT acquisition is very similar to planar gamma camera imaging, the same radiopharmaceuticals may be used. If a patient is examined in another type of nuclear medicine scan but the images are non-diagnostic, it may be possible to proceed straight to SPECT by moving the patient to a SPECT instrument, or even by simply reconfiguring the camera for SPECT image acquisition while the patient remains on the table.

To acquire SPECT images, the gamma camera is rotated around the patient. Projections are acquired at defined points during the rotation, typically every 3-6 degrees. In most cases, a full 360 degree rotation is used to obtain an optimal reconstruction. The time taken to obtain each projection is also variable, but 15-20 seconds is typical. This gives a total scan time of 15-20 minutes.

Multi-headed gamma cameras can provide accelerated acquisition. For example, a dual headed camera can be used with heads spaced 180 degrees apart, allowing 2 projections to be acquired simultaneously, with each head requiring 180 degrees of rotation. Triple-head cameras with 120 degree spacing are also used.

Cardiac gated acquisitions are possible with SPECT, just as with planar imaging techniques such as MUGA. Triggered by Electrocardiogram (EKG) to obtain differential information about the heart in various parts of its cycle, gated myocardial SPECT can be used to obtain quantitative information about myocardial perfusion, thickness, and contractility of the myocardium during various parts of the cardiac cycle; and also to allow calculation of left ventricular ejection fraction, stroke volume, and cardiac output.

Application

SPECT can be used to complement any gamma imaging study, where a true 3D representation can be helpful. E.g. tumor imaging, infection (leukocyte) imaging, thyroid imaging or bone imaging.

Because SPECT permits accurate localisation in 3D space, it can be used to provide information about localised function in internal organs, such as functional cardiac or brain imaging.

Myocardial Perfusion Imaging

Myocardial perfusion imaging (MPI) is a form of functional cardiac imaging, used for the diagnosis of ischemic heart disease. The underlying principle is that under conditions of stress, diseased myocardium receives less blood flow than normal myocardium. MPI is one of several types of cardiac stress test.

A cardiac specific radiopharmaceutical is administered. e.g. 99mTc-tetrofosmin (Myoview, GE healthcare), 99mTc-sestamibi (Cardiolite, Bristol-Myers Squibb). Following this, the heart rate is raised to induce myocardial stress, either by exercise or pharmacologically with adenosine, dobutamine or dipyridamole (aminophylline can be used to reverse the effects of dipyridamole).

SPECT imaging performed after stress reveals the distribution of the radiopharmaceutical, and therefore the relative blood flow to the different regions of the myocardium. Diagnosis is made by comparing stress images to a further set of images obtained at rest. As the radionuclide redistributes slowly, it is not usually possible to perform both sets of images on the same day, hence a second attendance is required 1-7 days later (although, with a Tl-201 myocardial perfusion study with dipyridamole, rest images can be acquired as little as two-hours post stress). However, if stress imaging is normal, it is unnecessary to perform rest imaging, as it too will be normal – thus stress imaging is normally performed

MPI has been demonstrated to have an overall accuracy of about 83% (sensitivity: 85%; specificity: 72%) and is comparable (or better) than other non-invasive tests for ischemic heart disease, including stress echocardiography.

Functional Brain Imaging

Usually the gamma-emitting tracer used in functional brain imaging is 99mTc-HMPAO (hexamethylpropylene amine oxime). ^{99m}Tc is a metastable nuclear isomer which emits gamma rays which can be detected by a gamma camera. When it is attached to HMPAO, this allows 99mTc to be taken up by brain tissue in a manner proportial to brain blood flow, in turn allowing brain blood flow to be assessed with the nuclear gamma camera.

POEP

Because blood flow in the brain is tightly coupled to local brain metabolism and energy use, the 99mTc-HMPAO tracer (as well as the similar ^{99m}Tc-EC tracer) is used to assess brain metabolism regionally, in an attempt to diagnose and differentiate the different causal pathologies of dementia. Meta analysis of many reported studies suggests that SPECT with this tracer is about 74% sensitive at diagnosing Alzheimer's disease, vs. 81% sensitivity for clinical exam (mental testing, etc.). More recent studies have show accuracy of SPECT in Alzheimer diagnosis as high as 88% PMID 16785801. In meta analysis, SPECT was superior to clinical exam and clinical criteria (91% vs. 70%) in being able to differentiate Alzheimer's disease from vascular dementias. PMID 15545324 This latter ability relates to SPECT's imaging of local metabolism of the brain, in which the patchy loss of cortical metabolism seen in multiple strokes differs clearly from the more even or "smooth" loss of non-occipital cortical brain function typical of Alzheimer's disease.

^{99m}Tc-HMPAO SPECT scanning competes with FDG PET scanning of the brain, which works to assess regional brain glucose metabolism, to provide very similar information about local brain damage from many processes. SPECT is more widely available, however, for the basic reason that the radioisotope generation technology is longer-lasting and far less expensive in SPECT, and the gamma scanning equipment is less expensive as well. The reason for this is that 99mTc (technetium-99m) is extracted from relatively simple technetium-99m generators which are delivered to hospitals and scanning centers weekly, to supply fresh radioisotope, whereas FDG PET relies on FDG which must be made in an expensive medical cyclotron and "hot-lab" (automated chemistry lab for radiopharmaceutical manufacture), then must be delivered directly to scanning sites, with delivery-fraction for each trip handicapped by its natural short 110 minute half-life.

Reconstruction

Reconstructed images typically have resolutions of 64 × 64 or 128 × 128 pixels, with the pixel sizes ranging from 3-6 mm. The number of projections acquired is chosen to be approximately equal to the width of the resulting images. In general, the resulting reconstructed images will be of lower resolution, have increased noise than planar images, and be susceptible to artifacts.

Scanning is time consuming, and it is essential that there is no patient movement during the scan time. Movement can cause significant degradation of the reconstructed images, although movement compensation reconstruction techniques can help with this. A highly uneven distribution of radiopharmaceutical also has the potential to cause artifacts. A very intense area of activity (e.g. the bladder) can cause extensive streaking of the images and obscure neighboring areas of activity. (This is a limitation of the filtered back projection reconstruction algorithm. Iterative reconstruction is an alternative algorithm which is growing in importance, as it is less sensitive to artifacts and can also correct for attenuation and depth dependent blurring).

Attenuation of the gamma rays within the patient can lead to significant underestimation of activity in deep tissues, compared to superficial tissues. Approximate correction is possible, based on relative position of the activity. However, optimal correction is obtained with measured attenuation values. Modern SPECT equipment is available with an integrated x-ray CT scanner. As X-ray CT images are an attenuation map of the tissues, this data can be incorporated into the SPECT reconstruction to correct for attenuation. It also provides a precisely registered CT image which can provide additional anatomical information.

ION

An ion is an atom or molecule which has lost or gained one or more electrons, giving it a positive or negative electrical charge. According to the Model of Bohr this will be from or in the outer shield 'n'.

A negatively charged ion, which has more electrons than it has protons, is known as an anion (Greek word *'ano'* means 'up'). Conversely, a positively-charged ion, which has fewer electrons than protons, is known as a cation.

An ion consisting of a single atom is called a monatomic ion, but if it consists of two or more atoms, it is a polyatomic ion. Polyatomic ions containing oxygen, such as carbonate and sulfate, are called oxyanions.

Ions are denoted in the same way as electrically neutral atoms and molecules except for the presence of a superscript indicating the sign of the net electric charge and the number of electrons lost or gained, if more than one. For example: H^+ and SO_4^{2-}.

Formation

Formation of Monatomic Ions

Monatomic ions are formed by the addition of electrons to the valence shell of the atom, which is the outer-most electron shell in an atom, or the losing of electrons from this shell. The inner shells of an atom are filled with electrons that are tightly bound to the positively charged atomic nucleus, and so do not participate in this kind of chemical interaction. The process of gaining or losing electrons from a neutral atom or molecule is called *ionization*.

Atoms can be ionized by bombardment with radiation, but the more usual process of ionization encountered in chemistry is the transfer of electrons between atoms or molecules. This transfer is usually driven by the attaining of stable ("closed shell") electronic configurations. Atoms will gain or lose electrons depending on which action takes the least energy.

For example, a sodium atom, Na, has a single electron in its valence shell, surrounding 2 stable, filled inner shells of 2 and 8 electrons. Since these filled shells are very stable, a sodium atom tends to lose its extra electron and attain this stable configuration, becoming a sodium *cation* in the process.

$$Na \rightarrow Na^+ + e^-$$

On the other hand, a chlorine atom, Cl has 7 electrons in its valence shell, which is one short of the stable, filled shell with 8 electrons. Thus, a chlorine atom tends to *gain* an extra electron and attain a stable 8-electron configuration, becoming a chloride *anion* in the process:

$Cl + e^- \rightarrow Cl^-$

This driving force is what causes sodium and chlorine to undergo a chemical reaction, where the "extra" electron is transferred from sodium to chlorine, forming sodium cations and chloride anions. Being oppositely charged, these cations and anions form ionic bonds and combine together to form sodium chloride, NaCl, more commonly known as rock salt.

$Na^+ + Cl- \rightarrow NaCl$

Polyatomic and molecular ions are often formed by the combination of elemental ions such as H^+ with neutral molecules or by the gain of such elemental ions from neutral molecules. A simple example of this is the ammonium ion NH_4^+ which can be formed by ammonia NH_3 accepting a proton, H^+. Ammonia and ammonium have the same number of electrons in essentially the same electronic configuration but differ in protons. The charge has been added by the addition of a proton (H^+) not the addition or removal of electrons. The distinction between this and the removal of an electron from the whole molecule is important in large systems because it usually results in much more stable ions with complete electron shells. For example NH_3^+ is not stable because of an incomplete valence shell around nitrogen and is in fact a radical ion.

Ionization Potential

The energy required to detach an electron in its lowest energy state from an atom or molecule of a gas with less net electric charge is called the ionization potential, or ionization energy. The nth ionization energy of an atom is the energy required to detach its nth electron after the first n - 1 electrons have already been detached.

Each successive ionization energy is markedly greater than the last. Particularly great increases occur after any given block of atomic orbitals is exhausted of electrons. For this reason, ions tend to form in ways that leave them with full orbital blocks. For example, sodium has one valence electron, in its outermost shell, so in ionized form it is commonly found with one lost electron, as Na+. On the other side of the periodic table, chlorine has seven valence electrons, so in ionized form it is commonly found with one gained electron, as Cl^-. Francium has the lowest ionization energy of all the elements and fluorine has the greatest. The ionization energy of metals is generally much lower than the ionization energy of nonmetals, which is why metals will generally lose electrons to form positively-charged ions while nonmetals will generally gain electrons to form negatively-charged ions.

A neutral atom contains an equal number of Z protons in the nucleus and Z electrons in the electron shell. The electrons' negative charges thus exactly cancel the protons' positive charges. In the simple view of the Free electron model, a passing electron is therefore not attracted to a neutral atom and cannot bind to it. In reality, however, the atomic electrons form a cloud into which the additional electron penetrates, thus being exposed to a net positive charge part of the time. Furthermore, the additional charge displaces the original electrons and all of the Z + 1 electrons rearrange into a new configuration.

Plasma (Physics)

A collection of non-aqueous gas-like ions, or even a gas containing a proportion of charged particles, is called a plasma, often called the fourth state of matter because its properties are quite different from solids, liquids, and gases. Astrophysical plasmas containing predominantly a mixture of electrons and protons, may make up as much as 99.9% of visible matter in the universe.

Applications

Ions are essential to life. Sodium, potassium, calcium and other ions play an important role in the cells of living organisms, particularly in cell membranes. They have many practical, everyday applications in items such as smoke detectors, and are also finding use in unconventional technologies such as ion engines. Inorganic dissolved ions are a component of total dissolved solids, an indicator of water quality in the world.

CHEMICAL BOND

A chemical bond is the physical process responsible for the attractive interactions between atoms and molecules, and that which confers stability to diatomic and polyatomic chemical compounds. The explanation of the attractive forces is a complex area that is described by the laws of quantum electrodynamics. In practice, however, chemists usually rely on quantum theory or qualitative descriptions that are less rigorous but more easily explained to describe chemical bonding. In general, strong chemical bonding is associated with the sharing or transfer of electrons between the participating atoms. Molecules, crystals, and diatomic gases—indeed most of the physical environment around us—are held together by chemical bonds, which dictate the structure of matter.

Bonds vary widely in their strength. Generally covalent and ionic bonds are often described as "strong", whereas hydrogen bonds and van der Waals' bonds are generally considered to be "weak". Care should be taken because the strongest of the "weak" bonds can be stronger than the weakest of the "strong" bonds.

Overview

Remembering that opposite charges attract, and that the electrons orbiting the nucleus are negatively charged and protons in the nucleus are positively charged, the most stable configuration of nuclei and electrons is one in which the electrons spend more time between nuclei, than anywhere else in space. These electrons cause the nuclei to be attracted to each other.

In the simplest view of a so-called covalent bond, one or more electrons (often a pair of electrons) are drawn into the space between the two atomic nuclei. Here the negatively charged electrons are attracted to the positive charges of both nuclei, instead of just their own. This overcomes the repulsion between the two positively charged nuclei of the two atoms, and so this overwhelming attraction holds the two nuclei in a relatively fixed configuration of equilibrium, even though they will still vibrate at equilibrium position. In summary, covalent bonding involves sharing of electrons in which the positively charged nuclei of two or more atoms simultaneously attract the negatively charged electrons that are being shared. In a polar covalent bond, one or more electrons are unequally shared between two nuclei.

In a simplified view of an ionic bond, the bonding electron is not shared at all, but transferred. In this type of bond, the outer atomic orbital of one atom has a vacancy which allows addition of one or more electrons. These newly added electrons potentially occupy a lower energy-state (effectively closer to more nuclear charge) than they experience in a different atom. Thus, one nucleus offers a more tightly-bound position to an electron than does another nucleus, with the result that one atom may transfer an electron to the other. This transfer causes one atom to assume a net positive charge, and the other to assume a net negative charge. The bond then results from electrostatic attraction between atoms, and the atoms become positive or negatively charged ions.

All bonds can be explained by quantum theory, but, in practice, simplification rules allow chemists to predict the strength, directionality, and polarity of bonds. The octet rule and VSEPR theory are two examples. More sophisticated theories are valence bond theory which includes orbital hybridization and resonance, and the linear combination of atomic orbitals molecular orbital method which includes ligand field theory. Electrostatics are used to describe bond polarities and the effects they have on chemical substances.

History

Early speculations into the nature of the chemical bond, from as early as the 12th century, supposed that certain types of chemical species were joined by a type of chemical affinity. In 1704, Issac Newton famously outlined his atomic bonding theory, in "Query 31" of his Opticks, whereby atoms attach to each other by some "force". Specifically, after acknowledging the various popular theories, in vogue at the time, of how atoms were reasoned to attach to each other, i.e. "hooked atoms", "glued together by rest", or "stuck together by conspiring motions", Newton states that he would rather infer from their cohesion, that:

"Particles attract one another by some force, which in immediate contact is exceedingly strong, at small distances performs the chemical operations, and reaches not far from the particles with any sensible effect."

In 1819, on the heels of the invention of the voltaic pile, Jöns Jakob Berzelius developed a theory of chemical combination stressing the electronegative and electropositive character of the combining atoms. By the mid 19th century, Edward Frankland, F.A. Kekule, A.S. Couper, A.M. Butlerov, and Hermann Kolbe, building on the theory of radicals, developed the theory of valency, originally called "combining power", in which compounds were joined owing to an attraction of positive and negative poles. In 1916, chemist Gilbert N. Lewis developed the concept of the electron-pair bond, in which two atoms may share one to six electrons, thus forming the single electron bond, a single bond, a double bond, or a triple bond:

In Lewis' own words:

"An electron may form a part of the shell of two different atoms and cannot be said to belong to either one exclusively."

That same year, Walther Kossel put forward a theory similar to Lewis' only his model assumed complete transfers of electrons between atoms, and was thus a model of ionic bonds. Both Lewis and Kossel structured their bonding models on that of Abegg's rule (1904).

In 1927, the first mathematically complete quantum description of a simple chemical bond, i.e. that produced by one electron in the hydrogen molecular ion, H_2^+, was derived by the Danish physicist Oyvind Burrau This work showed that the quantum approach to chemical bonds could be fundamentally and quantitatively correct, but the mathematical methods used could not be extended to molecules containing more than one electron. A more practical, albeit less quantitative, approach was put forward in the same year by Walter Heitler and Fritz London. The Heitler-London method forms the basis of what is now called valence bond theory. In 1929, the linear combination of atomic orbitals molecular orbital method (LCAO) approximation was introduced by Sir John Lennard-Jones, who also suggested methods to derive electronic structures of molecules of F_2 (fluorine) and O_2 (oxygen) molecules, from basic quantum principles. This molecular orbital theory represented a covalent bond as a orbitals formed by combining the quantum mechanical Schrödinger atomic orbitals which had been hypothesized for electrons in single atoms. The equations for bonding electrons in multi-electron atoms could not be solved to mathematical perfection (i.e., analytically), but approximations for them still gave many good qualitative preditions and results. Most quantitative calculations in modern quantum chemistry use either valence bond or molecular orbital theory as a starting point, although a third approach, Density Functional Theory, has become increasingly popular in recent years.

In 1935, H. H. James and A. S. Coolidge carried out a calculation on the dihydrogen molecule that, unlike all previous calculation which used functions only of the distance of the electron from the atomic nucleus, used functions which also explicitly added the distance between the two electrons. With up to 13 adjustable parameters they obtained a result very close to the experimental result for the dissociation energy. Later extensions have used up to 54 parameters and give excellent agreement with experiment. This calculation convinced the scientific community that quantum theory could give agreement with experiment. However this approach has none of the physical pictures of the valence bond and molecular orbital theories and is difficult to extend to larger molecules.

Valence Bond Theory

In the year 1927, valence bond theory was formulated which argued essentially that a chemical bond forms when two valence electrons, in their respective atomic orbitals, work or function to hold two nuclei together, by virtue of system energy lowering effects. In 1931, building on this theory, chemist Linus Pauling published what some consider one of the most important papers in the history of chemistry: "On the Nature of the Chemical Bond". In this paper, building on the works of Lewis, and the valence bond theory (VB) of Heitler and London, and his own earlier work, he presented six rules for the shared electron bond, the first three of which were already generally known:

1. The electron-pair bond forms through the interaction of an unpaired electron on each of two atoms.
2. The spins of the electrons have to be opposed.
3. Once paired, the two electrons cannot take part in additional bonds.

His last three rules were new:

1. The electron-exchange terms for the bond involves only one wave function from each atom.
2. The available electrons in the lowest energy level form the strongest bonds.
3. Of two orbitals in an atom, the one that can overlap the most with an orbital from another atom will form the strongest bond, and this bond will tend to lie in the direction of the concentrated orbital.

Building on this article, Pauling's 1939 textbook: On the Nature of the Chemical Bond would become what some have called the "bible" of modern chemistry. This book helped experimental chemists to understand the impact of quantum theory on chemistry. However, the later edition in 1959 failed to address adequately the problems that appeared to be better understood by molecular orbital theory. The impact of valence theory declined during the 1960's and 1970's as molecular orbital theory grew in popularity and was implemented in many large computer programs. Since the 1980s, the more difficult problems of implementing valence bond theory into computer programs have been largely solved and valence bond theory has seen a resurgence.

Molecular Orbital Theory

Molecular orbital theory (MO) uses a linear combination of atomic orbitals to form molecular orbitals which cover the whole molecule. These are often divided into bonding orbitals, anti-bonding orbitals, and non-bonding orbitals. A molecular orbital is merely a Schrödinger orbital which includes several, but often only two nuclei. If this orbital is of type in which the electron(s) in the orbital have a higher probability of being between nuclei than elsewhere, the orbital will be a bonding orbital, and will tend to hold the nuclei together. If the electrons tend to be present in a molecular orbital in which they spend more time elsewhere than between the nuclei, the orbital will function as an anti-bonding orbital and will actually weaken the bond. Electrons in non-bonding orbitals tend to be in deep orbitals (nearly atomic orbitals) associated almost entirely with one nucleus or the other, and thus they spend equal time between nuclei or not. These electrons neither contribute nor detract from bond strength.

Comparison of Valence Bond and Molecular Orbital Theory

In some respects valence bond theory is superior to molecular orbital theory. When applied to the simplest two-electron molecule, H_2, valence bond theory, even at the simplest Heitler-London approach, gives a much closer approximation to the bond energy, and it provides a much more accurate representation of the behavior of the electrons as chemical bonds are formed and broken. In contrast simple molecular orbital theory predicts that the hydrogen molecule dissociates into a linear superposition of hydrogen atoms and positive and negative hydrogen ions, a completely unphysical result. This explains in part why the curve of total energy against interatomic distance for the valence bond method lies above the curve for the molecular orbital method at all distances and most particularly so for large distances. This situation arises for all homonuclear diatomic molecules and is particularly a problem for F_2, where the minimum energy of the curve with molecular orbital theory is still higher in energy than the energy of two F atoms.

The concepts of hybridization are so versatile, and the variability in bonding in most organic compounds is so modest, that valence bond theory remains an integral part of the vocabulary of organic chemistry. However, the work of Friedrich Hund, Robert Mulliken, and Gerhard Herzberg showed that molecular orbital theory provided a more appropriate description of the spectroscopic, ionization and magnetic properties of molecules. The deficiencies of valence bond theory became apparent when hypervalent molecules (e.g. PF_5) were explained without the use of d orbitals that were crucial to the bonding hybridisation scheme proposed for such molecules by Pauling. Metal complexes and electron deficient compounds (e.g. diborane) also appeared to be well described by molecular orbital theory, although valence bond descriptions have been made.

In the 1930s the two methods strongly competed until it was realised that they are both approximations to a better theory. If we take the simple valence bond structure and mix in all possible covalent and ionic structures arising from a particular set of atomic orbitals, we reach what is called the full configuration interaction wave function. If we take the simple molecular orbital description of the ground state and combine that function with the functions describing all possible excited states using unoccupied orbitals arising from the same set of atomic orbitals, we also reach the full configuration interaction wavefunction. It can be then seen that the simple molecular orbital approach gives too much weight to the ionic structures, while the simple valence bond approach gives too little. This can also be described as saying that the molecular orbital approach is too delocalised, while the valence bond approach is too localised.

The two approaches are now regarded as complementary, each providing its own insights into the problem of chemical bonding. Modern calculations in quantum chemistry usually start from (but ultimately go far beyond) a molecular orbital rather than a valence bond approach, not because of any intrinsic superiority in the former but rather because the MO approach is more readily adapted to numerical computations. However better valence bond programs are now available.

Bonds in Chemical Formula

The 3-dimensionality of atoms and molecules makes it difficult to use a single technique for indicating orbitals and bonds. In molecular formulae the chemical bonds (binding orbitals) between atoms are indicated by various different methods according to the type of discussion. Sometimes, they are completely neglected. For example, in organic chemistry chemists are sometimes concerned only with the functional groups of the molecule. Thus, the molecular formula of ethanol (a compound in alcoholic beverages) may be written in a paper in conformational, 3-dimensional, full 2-dimensional (indicating every bond with no 3-dimensional directions), compressed 2-dimensional (CH_3–CH_2–OH), separating the functional group from another part of the molecule (C_2H_5OH), or by its atomic constituents (C_2H_6O), according to what is discussed. Sometimes, even the non-bonding valence shell electrons (with the 2-dimensional approximate directions) are marked, i.e. for elemental carbon 'C'. Some chemists may also mark the respective orbitals, i.e. the hypothetical ethene-4 anion, $C=C>^{-4}$ indicating the possibility of bond formation.

Strong Chemical Bonds

These chemical bonds are intramolecular forces, which hold atoms together in molecules. In the simplistic localized view of bonding, the number of electrons participating in a bond (or located in a bonding orbital) is typically multiples of two, four, or six, respectively. Even numbers are common because electrons enjoy lower energy states, if paired. Substantially more advanced bonding theories have shown that bond strength is not always a whole number, depending on the distribution of electrons to each atom involved in a bond. For example, the carbons in benzene are connected to each other with about 1.5 bonds, and the two atoms in nitric oxide NO, are connected with about 2.5 bonds. Quadruple bonds are also well known. The type of strong bond depends on the difference in electronegativity and the distribution of the electron orbital paths available to the atoms that are bonded. The larger the difference in electronegativity, the more an electron is attracted to a particular atom involved in the bond, and the more "ionic" properties the bond is said to have ("ionic" means the bond electron(s) are unequally shared). The smaller the difference in electronegativity, the more covalent properties (full sharing) the bond has.

Covalent Bond

Covalent bonding is a common type of bonding, in which the electronegativity difference between the bonded atoms is small or non-existent. Bonds within most organic compounds are described as covalent. See sigma bonds and pi bonds for LCAO-description of such bonding.

Polar Covalent Bond

Polar covalent bonding is intermediate in character between a covalent and an ionic bond.

Coordinate Covalent Bond

Ionic Bond

Ionic bonding is a type of electrostatic interaction between atoms which have a large electronegativity difference. There is no precise value that distinguishes ionic from covalent bonding but a difference of electronegativity of over 2.0 is likely to be ionic and a difference of less than 1.5 is likely to be covalent. Ionic bonding leads to separate positive and negative ions. Ionic charges are commonly between -3e to +3e.

Coordinate covalent bonding, sometimes referred to as dative bonding, is a kind of covalent bonding, in which the covalent bonding electrons originate solely from one of the atoms, the electron-pair donor or Lewis base but are approximately equally shared in the formation of a covalent bond. This concept is somewhat fading as chemists increasingly embrace molecular orbital theory. Examples of coordinate covalent bonding occur in nitrones and ammonia borane. The arrangement is different from an ionic bond in that the electronegativity difference is small, resulting in covalency. They are shown by an arrow unlike other bonds. This arrow shows its head towards the electron acceptor or lewis acid and its tail towards the lewis base. This bond is seen in ammonium.

One- and Three-electron Bonds

Bonds with one or three electrons can be found in radical species, which have an odd number of electrons. The simplest example of a 1-electron bond is found in the hydrogen molecular cation, $H_2^{+\cdot}$ One-electron bonds often have about half the bond energy of a 2-electron bond, and are therefore called "half bonds". However, there are exceptions: in the case of dilithium, the bond is actually stronger for the 1-electron Li_2^+ than for the 2-electron Li_2. This exception can be explained in terms of hybridization and inner-shell effects.

The simplest example of three-electron bonding can be found in the helium dimer cation, He_2^+, and can also be considered a "half bond" because, in molecular orbital terms, the third electron is in an anti-bonding orbital which cancels out half of the bond formed by the other two electrons. Another example of a molecule containing a 3-electron bond, in addition to two 2-electron bonds, is nitric oxide, NO. The oxygen molecule, O_2 can also be regarded as having two 3-electron bonds and one 2-electron bond, which accounts for its paramagnetism and its formal bond order of 2.

Molecules with odd-electron bonds are usually highly reactive. These types of bond are only stable between atoms with similar electronegativities.

Bent Bonds

Bent bonds, also known as banana bonds, are bonds in strained or otherwise sterically hindered molecules those binding orbitals are forced into a banana-like form. Bent bonds are often more susceptible to reactions than ordinary bonds.

3c-2e and 3c-4e Bonds

In three-center two-electron bonds ("3c-2e") three atoms share two electrons in bonding. This type of bonding occurs in electron deficient compounds like diborane. Each such bond (2 per molecule in diborane) contains a pair of electrons which connect the boron atoms to each other in a banana shape (shown as a more sharply angled section in the stick model at right), with a proton (nucleus of a hydrogen atom) in the middle of the bond, sharing electrons with both boron atoms.

Three-center four-electron bonds ("3c-4e") also exist which explain the bonding in hypervalent molecules. In certain cluster compounds, so-called four-center two-electron bonds also have been postulated.

In certain congugated π (pi) systems, such as benzene and other aromatic compounds and in congugated network solids such as graphite, the electrons in the congugated system of p-bonds are spread over as many nuclear centers as exist in the molecule or the network.

Aromatic Bond

In most cases, the locations of electrons cannot be simplified to simple lines (place for two electrons) or dots (a single electron). In aromatic bonds which occur in planar rings of atoms where the 4n+2 rule determines whether ring molecules would show extra stability.

In benzene, the prototypical aromatic compound, 18 bonding electrons bind 6 carbon atoms together to form a planar ring structure. The bond "order" (average number of bonds) between the different carbon atoms may be said to be (18/6)/2=1.5, but in this case the bonds are all identical from the chemical point of view. They may sometimes be written as single bonds alternating with double bonds, but the view of all ring bonds as being equivalently about 1.5 bonds in strength, is much closer to truth.

In the case of heterocyclic aromatics and substituted benzenes, the electronegativity differences between different parts of the ring may dominate the chemical behaviour of aromatic ring bonds, which otherwise are equivalent.

Metallic Bond

In a metallic bond, bonding electrons are delocalized over a lattice of atoms. By contrast, in ionic compounds, the locations of the binding electrons and their charges are static. Because of delocalization or the free moving of electrons, it leads to the metallic properties such as conductivity, ductility and hardness.

Intermolecular Bonding

There are four basic types of bonds that can be formed between two or more (otherwise non-associated) molecules, ions or atoms. Intermolecular forces cause molecules to be attracted or repulsed by each other. Often, these define some of the physical characteristics (such as the melting point) of a sub-permanent dipole to permanent dipole.

- A large electronegativity difference between two strongly bonded atoms within a molecule causes a dipole to form (a dipole is a pair of permanent partial charges). Dipoles will attract or repel each other.
- *Hydrogen Bond:* In some ways this is an especially strong example of a permanent dipole, as above. However, in the hydrogen bond, the hydrogen proton comes closer to being shared between target and donor atoms, in a three-center two-electron bond like that in diborane. Hydrogen bonds explain the relatively high boiling points of liquids like water, ammonia, and hydrogen fluoride, compared with their heavier counterparts in the same periodic table column.
- Instantaneous dipole to induced dipole (van der Waals) instantaneous dipole to induced dipole, or van der Waals forces, are the weakest, but also the most

prolific—occurring between all chemical substances. Imagine a helium atom: At any one point in time, the electron cloud around the (otherwise neutral) atom can be thought to be slightly imbalanced, with momentarily more negative charge on one side. This is referred to as an instantaneous dipole. This dipole, with its slight charge imbalance, may attract or repel the electrons within a neighbouring helium atom, setting up another dipole. The two atoms will be attracted for an instant, before the charge rebalances and the atoms move on.

Cation-pi Interaction: Cation-pi interactions occur between the localized negative charge of p orbital electrons, located above and below the plane of an aromatic ring, and a positive charge.

Electrons in Chemical Bonds

Many simple compounds involve covalent bonds. These molecules have structures that can be predicted using valence bond theory, and the properties of atoms involved can be understood using concepts such as oxidation number. Other compounds that involve ionic structures can be understood using theories from classical physics.

In the case of ionic bonding, electrons are mainly localized on the individual atoms, and electrons do not travel between the atoms very much. Each atom is assigned an overall electric charge to help conceptualize the molecular orbital's distribution. The forces between atoms (or ions) are largely characterized by isotropic continuum electrostatic potentials.

By contrast, in covalent bonding, the electron density within a bond is not assigned to individual atoms, but is instead delocalized in the MOs between atoms. The widely accepted theory of the linear combination of atomic orbitals (LCAO) helps describe the molecular orbital structures and energies based on the atomic orbitals of the atoms they came from. Unlike pure ionic bonds, covalent bonds may have directed anisotropic properties. These may have their own names, too, such as Sigma and Pi bond.

Atoms can also form bonds that are intermediates between ionic and covalent. This is because these definitions are based on the extent of electron delocalization. Electrons can be partially delocalized between atoms, but spend more time around one atom than another. This type of bond is often called polar covalent. See electronegativity.

Thus, the electrons in a molecular orbital (or 'in a polar covalent, or in a covalent bond') can be said to be either localized on certain atom(s) or delocalized between two or more atoms. The type of bond between two atoms is defined by how much the electron density is localized or delocalized among the atoms of the bonds.

CHAPTER 11 QUANTUM FIELD THEORY

Quantum field theory (QFT) provides a theoretical framework for constructing quantum mechanical models of systems classically described by fields or of many-body systems. It is widely used in particle physics and condensed matter physics. Most theories in modern particle physics, including the Standard Model of elementary particles and their interactions, are formulated as relativistic quantum field theories. In condensed matter physics, quantum field theories are used in many circumstances, especially those where the number of particles is allowed to fluctuate—for example, in the BCS theory of superconductivity.

In quantum field theory (QFT) the forces between particles are mediated by other particles. For instance, the electromagnetic force between two electrons is caused by an exchange of photons. But quantum field theory applies to all fundamental forces. Intermediate vector bosons mediate the weak force, gluons mediate the strong force, and gravitons mediate the gravitational force. These force carrying particles are virtual particles and, by definition, cannot be detected while carrying the force, because such detection will imply that the force is not being carried.

In QFT photons are not thought of as 'little billiard balls', they are considered to be field quanta - necessarily chunked ripples in a field that 'look like' particles. Fermions, like the electron, can also be described as ripples in a field, where each kind of fermion has its own field. In summary, the classical visualisation of "everything is particles and fields", in quantum field theory, resolves into "everything is particles", which then resolves into "everything is fields". In the end, particles are regarded as excited states of a field (field quanta).

History

Quantum field theory originated in the 1920s from the problem of creating a quantum mechanical theory of the electromagnetic field. In 1926, Max Born, Pascual Jordan, and Werner Heisenberg constructed such a theory by expressing the field's internal degrees of freedom as an infinite set of harmonic oscillators and by employing the usual procedure for quantizing those oscillators (canonical quantization). This theory assumed that no electric charges or currents were present and today would be called a free field theory. The first reasonably complete theory of quantum electrodynamics, which included both the electromagnetic field and electrically charged matter (specifically, electrons) as quantum mechanical objects, was created by Paul Dirac in 1927. This quantum field theory could be used to model important processes such as the emission of a photon by an electron dropping into a quantum state of lower energy, a process in which the number of particles changes — one atom in the initial state becomes an atom plus a photon in the final state. It is now understood that the ability to describe such processes is one of the most important features of quantum field theory.

It was evident from the beginning that a proper quantum treatment of the electromagnetic field had to somehow incorporate Einstein's relativity theory, which had after all grown out of the study of classical electromagnetism. This need to put together relativity and quantum mechanics was the second major motivation in the development of quantum field theory. Pascual Jordan and Wolfgang Pauli showed in 1928 that quantum fields could be made to behave in the way predicted by special relativity during coordinate transformations (specifically, they showed that the field commutators were Lorentz invariant), and in 1933 Niels Bohr and Leon Rosenfeld showed that this result could be interpreted as a limitation on the ability to measure fields at space-like separations, exactly as required by relativity. A further boost for quantum field theory came with the discovery of the Dirac equation, a single-particle equation obeying both relativity and quantum mechanics, when it was shown that several of its undesirable properties (such as negative-energy states) could be eliminated by reformulating the Dirac equation as a quantum field theory. This work was performed by Wendell Furry, Robert Oppenheimer, Vladimir Fock, and others.

The third thread in the development of quantum field theory was the need to handle the statistics of many-particle systems consistently and with ease. In 1927, Jordan tried to extend the canonical quantization of fields to the many-body wavefunctions of identical particles, a procedure that is sometimes called second quantization. In 1928, Jordan and Eugene Wigner found that the quantum field describing electrons, or other fermions, had to be expanded using anti-commuting creation and annihilation operators due to the Pauli exclusion principle. This thread of development was incorporated into many-body theory, and strongly influenced condensed matter physics and nuclear physics.

Despite its early successes, quantum field theory was plagued by several serious theoretical difficulties. Many seemingly-innocuous physical quantities, such as the energy shift of electron states due to the presence of the electromagnetic field, gave infinity—a nonsensical result—

when computed using quantum field theory. This "divergence problem" was solved during the 1940s by Bethe, Tomonaga, Schwinger, Feynman, and Dyson, through the procedure known as renormalization. This phase of development culminated with the construction of the modern theory of quantum electrodynamics (QED). Beginning in the 1950s with the work of Yang and Mills, QED was generalized to a class of quantum field theories known as gauge theories. The 1960s and 1970s saw the formulation of a gauge theory now known as the Standard Model of particle physics, which describes all known elementary particles and the interactions between them. The weak interaction part of the standard model was formulated by Sheldon Glashow, with the Higgs mechanism added by Steven Weinberg and Abdus Salam. The theory was shown to be consistent by Gerardus 't Hooft and Martinus Veltman.

Also during the 1970s, parallel developments in the study of phase transitions in condensed matter physics led Leo Kadanoff, Michael Fisher and Kenneth Wilson (extending work of Ernst Stueckelberg, Andre Peterman, Murray Gell-Mann and Francis Low) to a set of ideas and methods known as the renormalization group. By providing a better physical understanding of the renormalization procedure invented in the 1940s, the renormalization group sparked what has been called the "grand synthesis" of theoretical physics, uniting the quantum field theoretical techniques used in particle physics and condensed matter physics into a single theoretical framework.

The study of quantum field theory is alive and flourishing, as are applications of this method to many physical problems. It remains one of the most vital areas of theoretical physics today, providing a common language to many branches of physics.

Principles of Quantum Field Theory

Classical Fields and Quantum Fields

Quantum mechanics, in its most general formulation, is a theory of abstract operators (observables) acting on an abstract state space (Hilbert space), where the observables represent physically-observable quantities and the state space represents the possible states of the system under study. Furthermore, each observable corresponds, in a technical sense, to the classical idea of a degree of freedom. For instance, the fundamental observables associated with the motion of a single quantum mechanical particle are the position and momentum operators $\hat{x}$ and $\hat{\rho}$. Ordinary quantum mechanics deals with systems such as this, which possess a small set of degrees of freedom.

(It is important to note, at this point, that this article does not use the word "particle" in the context of wave–particle duality. In quantum field theory, "particle" is a generic term for any discrete quantum mechanical entity, such as an electron, which can behave like classical particles or classical waves under different experimental conditions.)

A **quantum field** is a quantum mechanical system containing a large, and possibly infinite, number of degrees of freedom. This is not as exotic a situation as one might think. A classical field contains a set of degrees of freedom at each point of space; for instance, the classical electromagnetic field defines two vectors—the electric field and the magnetic field—that can in principle take on distinct values for each position r. When the field as a whole is considered as a quantum mechanical system, its observables form an infinite (in fact uncountable) set, because r is continuous.

Furthermore, the degrees of freedom in a quantum field are arranged in "repeated" sets. For example, the degrees of freedom in an electromagnetic field can be grouped according to the position r, with exactly two vectors for each *r*. Note that r is an ordinary number that "indexes" the observables; it is not to be confused with the position operator $\hat{x}$ encountered in ordinary quantum mechanics, which is an observable. (Thus, ordinary quantum mechanics is sometimes referred to as "zero-dimensional quantum field theory", because it contains only a single set of observables.) It is also important to note that there is nothing special about r because, as it turns out, there is generally more than one way of indexing the degrees of freedom in the field.

In the following sections, we will show how these ideas can be used to construct a quantum mechanical theory with the desired properties. We will begin by discussing single-particle quantum mechanics and the associated theory of many-particle quantum mechanics. Then, by finding a way to index the degrees of freedom in the many-particle problem, we will construct a quantum field and study its implications.

Single-particle and many-particle Quantum Mechanics

In ordinary quantum mechanics, the time-dependent Schrödinger equation describing the time evolution of the quantum state of a single non-relativistic particle is

$$\frac{-\hbar^2}{2m}\frac{-\partial^2}{\partial x^2}\Psi(x, t) + V(x)\,\Psi(x, t) = i\hbar\frac{\partial}{\partial t}\Psi(x, t)$$

where m is the particle's mass, V is the applied potential, and (ψ) denotes the quantum state (we are using bracket notation).

We wish to consider how this problem generalizes to N particles. There are two motivations for studying the many-particle problem. The first is a straightforward need in condensed matter physics, where typically the number of particles is on the order of Avogadro's number

(6.0221415×10^{23}). The second motivation for the many-particle problem arises from particle physics and the desire to incorporate the effects of special relativity. If one attempts to include the relativistic rest energy into the above equation, the result is either the Klein-Gordon equation or the Dirac equation. However, these equations have many unsatisfactory qualities; for instance, they possess energy eigenvalues which extend to $-\infty$, so that there seems to be no easy definition of a ground state. It turns out that such inconsistencies arise from neglecting the possibility of dynamically creating or destroying particles, which is a crucial aspect of relativity. Einstein's famous mass-energy relation predicts that sufficiently massive particles can decay into several lighter particles, and sufficiently energetic particles can combine to form massive particles. For example, an electron and a positron can annihilate each other to create photons. Thus, a consistent relativistic quantum theory must be formulated as a many-particle theory.

Furthermore, we will assume that the N particles are indistinguishable. As described in the article on identical particles, this implies that the state of the entire system must be either symmetric (bosons) or antisymmetric (fermions) when the coordinates of its constituent particles are exchanged. These multi-particle states are rather complicated to write. For example, the general quantum state of a system of N bosons is written as

$$|\phi 1 \ldots \phi N\rangle = \sqrt{\frac{\pi_j N_j!}{N_j}} \sum_{p \varepsilon S_N} |\phi p(1)\rangle \ldots |\phi p(N)\rangle$$

where $|\phi_i\rangle$ are the single-particle states, N_j is the number of particles occupying state j, and the sum is taken over all possible permutations p acting on N elements. In general, this is a sum of $N!$ (N factorial) distinct terms, which quickly becomes unmanageable as N increases. The way to simplify this problem is to turn it into a quantum field theory.

Second Quantization

In this section, we will describe a method for constructing a quantum field theory called second quantization. This basically involves choosing a way to index the quantum mechanical degrees of freedom in the space of multiple identical-particle states. It is based on the Hamiltonian formulation of quantum mechanics; several other approaches exist, such as the Feynman path integral , which uses a Lagrangian formulation. For an overview, see the article on quantization.

Second Quantization of Bosons

For simplicity, we will first discuss second quantization for bosons, which form perfectly symmetric quantum states. Let us denote the mutually orthogonal single-particle states by

$$|\phi_1\rangle, |\phi_2\rangle, |\phi_3\rangle$$

and so on. For example, the 3-particle state with one particle in state

$$|\phi_1\rangle$$

and two in state

$|\phi_2\rangle$ is

$$\frac{1}{\sqrt{3}}|\phi_1\rangle|\phi_2\rangle|\phi_2\rangle + |\phi_2\rangle|\phi_1\rangle|\phi_2\rangle + |\phi_2\rangle|\phi_2\rangle|\phi_1\rangle.$$

The first step in second quantization is to express such quantum states in terms of **occupation numbers,** by listing the number of particles occupying each of the single-particle states

$|\phi_1\rangle, |\phi_2\rangle$ etc.

This is simply another way of labelling the states. For instance, the above 3-particle state is denoted as $|1, 2, 0, 0, 0, \ldots\rangle$.

The next step is to expand the *N*-particle state space to include the state spaces for all possible values of *N*. This extended state space, known as a Fock space, is composed of the state space of a system with no particles (the so-called vacuum state), plus the state space of a 1-particle system, plus the state space of a 2-particle system, and so forth. It is easy to see that there is a one-to-one correspondence between the occupation number representation and valid boson states in the Fock space.

At this point, the quantum mechanical system has become a quantum field in the sense we described above. The field's elementary degrees of freedom are the occupation numbers, and each occupation number is indexed by a number *j* ... indicating which of the single-particle states $|\phi_1\rangle |\phi_2\rangle, \ldots |\phi_j\rangle \ldots$ it refers to.

The properties of this quantum field can be explored by defining creation and annihilation operators, which add and subtract particles. They are analogous to "ladder operators" in the quantum harmonic oscillator problem, which added and subtracted energy quanta. However, these operators literally create and annihilate particles of a given quantum state. The bosonic annihilation operator a_2 and creation operator have $a^{\dagger}_2$ the following effects:

$$a_2 |N_1, N_2, N_3, \ldots\rangle = \sqrt{N_2}\, |N_1, (N_2 - 1), N_3, \ldots\rangle$$

$$a^{\dagger}_2 |N_1, N_2, N_3, \ldots\rangle = \sqrt{N_2+1}\, |N_1, (N_2 + 1), N_3, \ldots\rangle$$

It can be shown that these are operators in the usual quantum mechanical sense, i.e. linear operators acting on the Fock space. Furthermore, they are indeed Hermitian conjugates, which justifies the way we have written them. They can be shown to obey the commutation relation

$$[a_1, a_j] = 0, [a^\dagger_i, a^\dagger_j] = 0, [a_i, a^\dagger_i] = \delta_{ij}$$

where d stands for the Kronecker delta. These are precisely the relations obeyed by the ladder operators for an infinite set of independent quantum harmonic oscillators, one for each single-particle state. Adding or removing bosons from each state is therefore analogous to exciting or de-exciting a quantum of energy in a harmonic oscillator.

The Hamiltonian of the quantum field (which, through the Schrödinger equation, determines its dynamics) can be written in terms of creation and annihilation operators. For instance, the Hamiltonian of a field of free (non-interacting) bosons is

$$H = \sum_k E_k a^\dagger_k a_k$$

where Ek is the energy of the k-th single-particle energy eigenstate. Note that

$$a^\dagger_k a_k | \ldots, N_k, \ldots\rangle = N_k | \ldots, N_k, \ldots\rangle$$

Second Quantization of Fermions

It turns out that a different definition of creation and annihilation must be used for describing fermions. According to the Pauli exclusion principle, fermions cannot share quantum states, so their occupation numbers Ni can only take on the value 0 or 1. The fermionic annihilation operators c and creation operators $c^\dagger$ are defined by

$$c_j | N_1, N_2, \ldots, N_j = 0, \ldots\rangle = 0$$

$$c_j | N_1, N_2, \ldots, N_j = 1, \ldots\rangle = (-1)^{(N_1 + \ldots N_j - i)} | N_1, N_2, \ldots, N_j = 0, \ldots\rangle$$

$$c^\dagger_j | N_1, N_2, \ldots, N_j = 0, \ldots\rangle = (-1)^{(N_1 + \ldots N_j - i)} | N_1, N_2, \ldots, N_j = 1, \ldots\rangle$$

$$c^\dagger_2 | N_1, N_2, \ldots N_j = 1, \ldots\rangle = 0$$

These obey an anticommutation relation:

$$\{ci, cj\} = 0, \{c^{+}{}_{i}\} = 0, \{c_i, c^{+}{}_{j}\} = \partial_{ij}$$

One may notice from this that applying a fermionic creation operator twice gives zero, so it is impossible for the particles to share single-particle states, in accordance with the exclusion principle.

Field Operators

We have previously mentioned that there can be more than one way of indexing the degrees of freedom in a quantum field. Second quantization indexes the field by enumerating the single-particle quantum states. However, as we have discussed, it is more natural to think about a "field", such as the electromagnetic field, as a set of degrees of freedom indexed by position.

To this end, we can define field operators that create or destroy a particle at a particular point in space. In particle physics, these operators turn out to be more convenient to work with, because they make it easier to formulate theories that satisfy the demands of relativity.

Single-particle states are usually enumerated in terms of their momenta (as in the particle in a box problem.) We can construct field operators by applying the Fourier transform to the creation and annihilation operators for these states. For example, the bosonic field annihilation operator ϕ (r) is

$$\phi(\mathrm{r}) \stackrel{def}{=} \sum_{j} e^{ikj.r} a_j$$

The bosonic field operators obey the commutation relation

$$[\phi(r), \phi(r')] = 0, [\phi^{\dagger}(r), \phi^{\dagger}(r')] = 0, [\phi(r), \phi^{\dagger}(r')] = \partial^3 (r-r')$$

where δ(x) stands for the Dirac delta function. As before, the fermionic relations are the same, with the commutators replaced by anticommutators.

It should be emphasized that the field operator is not the same thing as a single-particle wavefunction. The former is an operator acting on the Fock space, and the latter is just a scalar field. However, they are closely related, and are indeed commonly denoted with the same symbol. If we have a Hamiltonian with a space representation, say

$$H = -\frac{\hbar^2}{2m}\sum_i \nabla^2_i + \sum_{i<j} U(|r_i - r_j|)$$

where the indices i and j run over all particles, then the field theory Hamiltonian is

$$H = -\frac{\hbar^2}{2m}\int d^3r\phi^{\dagger}(r)\nabla^2\phi(r) + \int d^3r \int d^3r'\phi^{\dagger}(r)\phi^{\dagger}(r')U(|r-r'|)\phi(r')\phi(r)$$

This looks remarkably like an expression for the expectation value of the energy, with *f* playing the role of the wavefunction. This relationship between the field operators and wavefunctions makes it very easy to formulate field theories starting from space-projected Hamiltonians.

Implications of Quantum Field Theory

Unification of Fields and Particles

The "second quantization" procedure that we have outlined in the previous section takes a set of single-particle quantum states as a starting point. Sometimes, it is impossible to define such single-particle states, and one must proceed directly to quantum field theory. For example, a quantum theory of the electromagnetic field must be a quantum field theory, because it is impossible (for various reasons) to define a wavefunction for a single photon. In such situations, the quantum field theory can be constructed by examining the mechanical properties of the classical field and guessing the corresponding quantum theory. The quantum field theories obtained in this way have the same properties as those obtained using second quantization, such as well-defined creation and annihilation operators obeying commutation or anticommutation relations.

Quantum field theory thus provides a unified framework for describing "field-like" objects (such as the electromagnetic field, whose excitations are photons) and "particle-like" objects (such as electrons, which are treated as excitations of an underlying electron field).

Physical meaning of Particle Indistinguishability

The second quantization procedure relies crucially on the particles being identical. We would not have been able to construct a quantum field theory from a distinguishable many-particle system, because there would have been no way of separating and indexing the degrees of freedom.

Many physicists prefer to take the converse interpretation, which is that quantum field theory explains what identical particles are. In ordinary quantum mechanics, there is not much theoretical motivation for using symmetric (bosonic) or antisymmetric (fermionic) states, and the need for such states is simply regarded as an empirical fact. From the point of view of quantum field theory, particles are identical if and only if they are excitations of the same underlying quantum field. Thus, the question "why are all electrons identical?" arises from mistakenly regarding individual electrons as fundamental objects, when in fact it is only the electron field that is fundamental.

Particle Conservation and Non-conservation

During second quantization, we started with a Hamiltonian and state space describing a fixed number of particles (N), and ended with a Hamiltonian and state space for an arbitrary number of particles. Of course, in many common situations N is an important and perfectly well-defined quantity, e.g. if we are describing a gas of atoms sealed in a box. From the point of view of quantum field theory, such situations are described by quantum states that are eigenstates of the number operator $\hat{N}$ which measures the total number of particles present. As with any quantum mechanical observable, $\hat{N}$ is conserved if it commutes with the Hamiltonian. In that case, the quantum state is trapped in the N-particle subspace of the total Fock space, and the situation could equally well be described by ordinary $\hat{N}$ particle quantum mechanics.

For example, we can see that the free-boson Hamiltonian described above conserves particle number. Whenever the Hamiltonian operates on a state, each particle destroyed by an annihilation operator a_k is immediately put back by the creation operator $a^{\dagger}_{k}$.

On the other hand, it is possible, and indeed common, to encounter quantum states that are not eigenstates of $\hat{N}$ which do not have well-defined particle numbers. Such states are difficult or impossible to handle using ordinary quantum mechanics, but they can be easily described in quantum field theory as quantum superpositions of states having different values of N. For example, suppose we have a bosonic field whose particles can be created or destroyed by interactions with a fermionic field. The Hamiltonian of the combined system would be given by the Hamiltonians of the free boson and free fermion fields, plus a "potential energy" term such as:

$$HI = \sum_{k,\, q} V_q \,(aq + a^{\dagger}_{-q})\, c^{\dagger}_{k+q}\, C_k$$

where

$a^{\dagger}_k$ and ak denotes the bosonic creation and annihilation operators

$c^{\dagger}_k$ and c_k denotes the fermionic creation and annihilation operators; and

Vq is a parameter that describes the strength of the interaction.

This "interaction term" describes processes in which a fermion in state k either absorbs or emits a boson, thereby being kicked into a different eigenstate k + q. (In fact, this type of Hamiltonian is used to describe interaction between conduction electrons and phonons in metals. The interaction between electrons and photons is treated in a similar way, but is a little more complicated because the role of spin must be taken into account.) One thing to notice here is that even if we start out with a fixed number of bosons, we will typically end up with a superposition of states with different numbers of bosons at later times. The number of fermions, however, is conserved in this case.

In condensed matter physics, states with ill-defined particle numbers are particularly important for describing the various superfluids. Many of the defining characteristics of a superfluid arise from the notion that its quantum state is a superposition of states with different particle numbers.

Axiomatic Approaches

The preceding description of quantum field theory follows the spirit in which most physicists approach the subject. However, it is not mathematically rigorous. Over the past several decades, there have been many attempts to put quantum field theory on a firm mathematical footing by formulating a set of axioms for it. These attempts fall into two broad classes.

The first class of axioms, first proposed during the 1950s, include the Wightman, Osterwalder-Schrader, and Haag-Kastler systems. They attempted to formalize the physicists' notion of an "operator-valued field" within the context of functional analysis, and enjoyed limited success. It was possible to prove that any quantum field theory satisfying these axioms satisfied certain general theorems, such as the spin-statistics theorem and the CPT theorem. Unfortunately, it proved extraordinarily difficult to show that any realistic field theory, including the Standard Model, satisfied these axioms. Most of the theories that could be treated with these analytic axioms were physically trivial, being restricted to low-dimensions and lacking interesting dynamics. The construction of theories satisfying one of these sets of axioms falls in the field of constructive quantum field theory. Important work was done in this area in the 1970s by Segal, Glimm, Jaffe and others.

During the 1980s, a second set of axioms based on geometric ideas was proposed. This line of investigation, which restricts its attention to a particular class of quantum field theories known as topological quantum field theories, is associated most closely with Michael Atiyah and Graeme Segal, and was notably expanded upon by Edward Witten, Richard Borcherds, and Maxim Kontsevich. However, most physically-relevant quantum field theories, such as the Standard Model, are not topological quantum field theories; the quantum field theory of the fractional quantum Hall effect is a notable exception. The main impact of axiomatic topological quantum field theory has been on mathematics, with important applications in representation theory, algebraic topology, and differential geometry.

Finding the proper axioms for quantum field theory is still an open and difficult problem in mathematics. One of the Millennium Prize Problems—proving the existence of a mass gap in Yang-Mills theory—is linked to this issue.

Phenomena Associated with Quantum Field Theory

In the previous part of the article, we described the most general properties of quantum field theories. Some of the quantum field theories studied in various fields of theoretical physics possess additional special properties, such as renormalizability, gauge symmetry, and supersymmetry. These are described in the following sections.

Early in the history of quantum field theory, it was found that many seemingly innocuous calculations, such as the perturbative shift in the energy of an electron due to the presence of the electromagnetic field, give infinite results. The reason is that the perturbation theory for the shift in an energy involves a sum over all other energy levels, and there are infinitely many levels at short distances which each give a finite contribution.

Many of these problems are related to failures in classical electrodynamics that were identified but unsolved in the 19th century, and they basically stem from the fact that many of the supposedly "intrinsic" properties of an electron are tied to the electromagnetic field which it carries around with it. The energy carried by a single electron—its self energy—is not simply the bare value, but also includes the energy contained in its electromagnetic field, its attendant cloud of photons. The energy in a field of a spherical source diverges in both classical and quantum mechanics, but as discovered by Weisskopf, in quantum mechanics the divergence is much milder, going only as the logarithm of the radius of the sphere.

The solution to the problem, presciently suggested by Stueckelberg, independently by Bethe after the crucial experiment by Lamb, implemented at one loop by Schwinger, and systematically extended to all loops by Feynman and Dyson, with converging work by Tomonaga in isolated postwar Japan, is called renormalization. The technique of renormalization recognizes that the problem is essentially purely mathematical, that extremely short distances are at fault. In order to define a theory on a continuum, first place a cutoff on the fields, by postulating that quanta cannot have energies above some extremely high value. This has the effect of replacing continuous space by a structure where very short wavelengths do not exist, as on a lattice. Lattices break rotational symmetry, and one of the crucial contributions made by Feynman, Pauli and Villars, and modernized by 't Hooft and Veltman, is a symmetry preserving cutoff for perturbation theory. There is no known symmetrical cutoff outside of perturbation theory, so for rigorous or numerical work people often use an actual lattice.

On a lattice, every quantity is finite but depends on the spacing. When taking the limit of zero spacing, we make sure that the physically-observable quantities like the observed electron mass stay fixed, which means that the constants in the Lagrangian defining the theory depend on the spacing. Hopefully, by allowing the constants to vary with the lattice spacing, all the results at long distances become insensitive to the lattice, defining a continuum limit.

The renormalization procedure only works for a certain class of quantum field theories, called renormalizable quantum field theories. A theory is perturbatively renormalizable when the constants in the Lagrangian only diverge at worst as logarithms of the lattice spacing for very short spacings. The continuum limit is then well defined in perturbation theory, and even if it is not fully well defined non-perturbatively, the problems only show up at distance scales which are exponentially small in the inverse coupling for weak couplings. The Standard Model of particle physics is perturbatively renormalizable, and so are its component theories (quantum electrodynamics/electroweak theory and quantum chromodynamics). Of the three components, quantum electrodynamics is believed to not have a continuum limit, while the asymptotically free SU(2) and SU(3) weak hypercharge and strong color interactions are nonperturbatively well defined.

The renormalization group describes how renormalizable theories emerge as the long distance low-energy effective field theory for any given high-energy theory. Because of this, renormalizable theories are insensitive to the precise nature of the underlying high-energy short-distance phenomena. This is a blessing because it allows physicists to formulate low energy theories without knowing the details of high energy phenomenon. It is also a curse, because once a renormalizable theory like the standard model is found to work, it gives very few clues to higher energy processes. The only way high energy processes can be seen in the standard model is when they allow otherwise forbidden events, or if they predict quantitative relations between the coupling constants.

Gauge Freedom

A gauge theory is a theory that admits a symmetry with a local parameter. For example, in every quantum theory the global phase of the wave function is arbitrary and does not represent something physical. Consequently, the theory is invariant under a global change of phases (adding a constant to the phase of all wave functions, everywhere); this is a global symmetry. In quantum electrodynamics, the theory is also invariant under a local change of phase, that is - one may shift the phase of all wave functions so that the shift may be different at every point in space-time. This is a local symmetry. However, in order for a well-defined derivative operator to exist, one must introduce a new field, the gauge field, which also transforms in order for the local change of variables (the phase in our example) not to affect the derivative. In quantum electrodynamics this gauge field is the electromagnetic field. The change of local gauge of variables is termed gauge transformation.

In quantum field theory the excitations of fields represent particles. The particle associated with excitations of the gauge field is the gauge boson, which is the photon in the case of quantum electrodynamics.

The degrees of freedom in quantum field theory are local fluctuations of the fields. The existence of a gauge symmetry reduces the number of degrees of freedom, simply because some fluctuations of the fields can be transformed to zero by gauge transformations, so they are equivalent to having no fluctuations at all, and they therefore have no physical meaning. Such fluctuations are usually called "non-physical degrees of freedom" or gauge artifacts; usually some of them have a negative norm, making them inadequate for a consistent theory. Therefore, if a classical field theory has a gauge symmetry, then its quantized version (i.e. the corresponding quantum field theory) will have this symmetry as well. In other words, a gauge symmetry cannot have a quantum anomaly. If a gauge symmetry is anomalous (i.e. not kept in the quantum theory) then the theory is non-consistent: for example, in quantum electrodynamics, had there been a gauge anomaly, this would require the appearance of photons with longitudinal polarization and polarization in the time direction, the latter having a negative norm, rendering the theory inconsistent; another possibility would be for these photons to appear only in intermediate processes but not in the final products of any interaction, making the theory non unitary and again inconsistent (see optical theorem).

In general, the gauge transformations of a theory consist several different transformations, which may not be commutative. These transformations are together described by a mathematical object known as a gauge group. Infinitesimal gauge transformations are the gauge group generators. Therefore the number of gauge bosons is the group dimension (i.e. number of generators forming a basis).

All the fundamental interactions in nature are described by gauge theories. These are:

- *Quantum electrodynamics,* whose gauge transformation is a local change of phase, so that the gauge group is U(1). The gauge boson is the photon.
- *Quantum chromodynamics,* whose gauge group is SU(3). The gauge bosons are eight gluons.
- The electroweak theory, whose gauge group is U (1) × SU (2) (a direct product of U (1) and SU (2).
- Gravity, whose classical theory is general relativity, admits the equivalence principle which is a form of gauge symmetry.

Supersymmetry

Supersymmetry assumes that every fundamental fermion has a superpartner that is a boson and vice versa. It was introduced in order to solve the so-called Hierarchy Problem, that is, to explain why particles not protected by any symmetry (like the Higgs boson) do not receive radiative corrections to its mass driving it to the larger scales (GUT, Planck...). It was soon realized that supersymmetry has other interesting properties: its gauged version is an extension of general relativity (Supergravity), and it is a key ingredient for the consistency of string theory.

The way supersymmetry protects the hierarchies is the following: since for every particle there is a superpartner with the same mass, any loop in a radiative correction is cancelled by the loop corresponding to its superpartner, rendering the theory UV finite.

CHAPTER 12

LIGHT-DEPENDENT REACTIONS

The light-dependent reactions, or light reactions, are the first stage of photosynthesis. In this process light energy is converted into chemical energy, in the form of the energy-carriers ATP and NADPH. In the light-independent reactions, or dark reactions, the formed NADPH and ATP drive the reduction of CO_2 to more useful organic compounds, such as glucose.

The light-dependent reactions take place on the thylakoid membrane inside a chloroplast. The inside of the thylakoid membrane is called the lumen, and outside the thylakoid membrane is the stroma, where the light-dependent take place. The thylakoid membrane contain some integral membrane protein complexes which catalyze the light reactions. There are four major protein complexes in the thylakoid membrane: Photosystem I (PSI), Photosystem II (PSII), Cytochrome b6f complex and ATP synthase. These four complexes work together to ultimately create the products ATP and NADPH.

The two photosystems absorb light energy through proteins containing pigments, such as chlorophyll. The lightdepedent reactions begins in photosystem II. When a special chlorophyll molecule of PSII absorbs a photon, an electron in this molecule becomes in higher energy level. Because this state of an electron is very unstable, the electron is transferred from one to another molecule creating an chain of redox reactions, called an electron transport chain (ETC). The electron flow goes from PSII to cytochrome b6f to PSI. In PSI the electron gets the energy from another photon. The final electron acceptor is NADPH. In oxygenic photosynthesis, the first electron donor is water, creating oxygen as wasteproduct. In anoxygenic photosynthesis various electron donors are used.

Cytochrome b6f and ATP synthase are working together to create ATP. This process is called photophosphorylation, which occurs in two different ways. In non-cyclic photophosphorylation, cytochrome b6f uses the energy of electrons from PSII to pump protons from the stroma to the lumen. The proton gradient across the thylakoid membrane creates a proton-motive force, used by ATP synthase to form ATP. In cyclic photophosphorylation, cytochrome b6f uses the energy of electrons from not only PSII, but also PSI to create more ATP and to stop the production of NADPH. Cyclic phosphorylation is important to create NADPH and ATP in the right proportion for the light-independent reactions.

The net-reaction of all light-dependent reactions in oxygenic photosynthesis is:

$$2H_2O + 2NADP^+ + 3ADP + 3P_i \rightarrow O_2 + 2NADPH + 3ATP$$

Light to Chemical Energy

The two photosystems are the protein complexes that absorb photons and are able to use this energy to create an electron transport chain. Photosystem I and II are very similar qua structure and function. They use special proteins, called light-harvesting complexes, to absorb the photons with very high effectiveness. If a special pigment molecule in a so-called reaction center absorbs a photon, an electron in this pigment becomes in the excited state and then leaves to another molecule in the reaction center. This reaction, called photoinduced charge separation is the start of the electron flow and is very unique because it converts light energy into chemical forms.

The Light-harvesting System

A common misconception is that photosynthesis relies only on chlorophyll pigments. The truth is that photosynthesis would be rather inefficient using only chlorophyll molecules. Chlorophyll molecules absorb light only at specific wavelengths. A large gap is present in the middle of the visible regions between approximately 450 and 650 nm. This gap corresponds to the peak of the solar spectrum, so failure to collect this light would constitute a considerable lost opportunity. That's why photosynthesis organisms had developed a light-harvesting system, which bundles different pigments to create a much wider absorption spectrum.

The light harvesting-system is composed of numerous light-harvesting complexes that completely surround the reaction center where the photoinduced charge separation takes place. Chlorophyl, carotenes and xanthophylls are arranged in such light-harvesting complexes or LHC-proteins. These pigments are referred to as accessory pigments and funnel the energy to a special pigment in the reaction center of PSI or PSII.

If a pigment molecule absorbs a photon, an electron in the molecule becomes excited. For most compounds that absorb light, the electron simply returns to the ground state and the absorbed energy is converted into heat. But in a LHC-protein the pigments are so arranged that the excitation energy can be transferred from one molecule to a nearby molecule. The rate of this process, called resonance energy transfer, depends strongly on the distance between the energy donor and energy acceptor molecules. For reasons of conservation of energy, energy transfer must be from a donor in the excited state to an acceptor of equal or lower energy. If the energy of a photon becomes lower, the wavelength becomes longer. The pigments in an LHC-protein are so arranged that pigments are very close to each other and that a pigment is near another pigment that absorbs photons with a longer wavelength. Consequently, the pigment in the reaction center has to absorb photons with the longest wavelength and can not transfer his energy to another pigment. The function of LHC-proteins is to create an constant supply of excitation-energy to the reaction center pigment. Every reaction center has a couple of LHC-proteins.

The Reaction Centre

This pigment molecule in the reaction center lie near the periplasmic side of the membrane. It is a dimer of two chlorophyll molecules. This dimer is called a special pair because of its fundamental role in photosynthesis. This special pair is slightly different in PSI and PSII reaction center. In PSII it absorbs photons with a wavelength of 680 nm and it is therefore called P680. In PSI it absorbs phtons at 700 nm and it is called P700. In bacteria the special pair is called P760, P840, P870 or P960.

If an electron of the special pair in the reaction center becomes excited, it can not transfer his energy to antoher pigment using resonance energy transfer. In normal cicrumstances the electron should return to the ground state, but because the reaction center is so arranged that a suitable electron acceptor is nearby, the excited electron can move from the initial molecule to the acceptor. This process results in the formation of a positive charge on the special pair (due to the loss of an electron) and a negative charge on the acceptor and is, hence, referred to as photoinduced charge separation.

This initial charge separation occurs in less than 10 picoseconds (10-11 seconds). In their high-energy states, the special pigment and the acceptor could undergo charge recombination; that is, the electron on the acceptor could move back to neutralize the positive charge on the special pair. Its return to the special pair would waste a valuable high-energy electron and simply convert the absorbed light energy into heat. Three factors in the structure of the reaction center work together to suppress charge recombination nearly completely.

- Another electron acceptor is less than 10 Å away from the first acceptor, and so the electron is rapidly transferred farther away from the special pair.
- An electron donor is less than 10 Å away from the special pair, and so the positive charge is neutralized by the transfer of another electron.
- The electron transfer from the electron acceptor to the positively charged special pair is especially slow: the transfer is so thermodynamically favorable that it takes place in the inverted region where electron-transfer rates become slower.

Thus, electron transfer proceeds effeciently from the first electron acceptor to the next, creating an electron transport chain that ends if it has reached NADPH.

Photosynthetic Electron Transport Chains in Chloroplasts

The photosynthesis process in chloroplasts begins when an electron of P680 of PSII becomes in an higher-energy level. This energy is used to reduce a chain of electron acceptors which have subsequently lowered redox-potentials. This chain of electron acceptors is known as a electron transport chain. When this chain reach PS I, an electron is again excited, creating a high redox-potential. The electron transport chain of photosynthesis is often put in an diagram called the z-scheme, because the redox diagram from P680 to P700 resembles the letter z.

The final product of PSII is plastoquinol, a mobile electron carrier in the membrane. Plastoquinol transfers the electron from PSII to the proton pump, cytochrome b6f. The ultimate electron donor of PSII is water. Cytochrome b6f proceeds the electron chain to PSI through plastocyanin molecules. PSI is able to continue the electron transfer in two different ways. It can transfer the electrons either to plastoquinonol again, creating a cyclic electron flow, or to an enzyme called FNR creating an non-cyclic electron flow. PSI releases FNR into the stroma, where it reduces NADP+ to NADPH.

Activities of the electron transport chain, especially from cytochrome b6f lead to pumping of protons from the stroma to the lumen. The resulting transmembrane proton gradient is used to make ATP via ATP synthase.

The overall process of the photosynthetic electron transport chain in chloroplasts is:

- H_2O →PS II - plastoquinone - cytb_6f - plastocyanin - PS I - NADPH

Photosystem II

PS II is an extremely complex, highly organized transmembrane structure that contains a water-splitting complex, chlorophylls and carotenoid pigments, a reaction center (P680), pheophytin (a pigment similar to chlorophyll), and two quinones. It uses the energy of sunlight to transfer electrons from water to a mobile electron carrier in the membrane called plastoquinone:

H_2O →P680 →P680* – plastoquinone

Plastoquinone, in turn, transfers electrons to b_6f, which feeds them into PS I.

The Water-splitting Complex

The step H_2O →P680 is performed by a poorly-understood structure embedded within PS II called the *water-*

splitting complex or the *oxygen-evolving complex*. It catalyzes a reaction that splits water into electrons, protons and oxygen:

$$2H_2O \rightarrow 4H^+ + 4e^- + O_2$$

The electrons are transferred to special chlorophyll molecules (embedded in PS II) that are promoted to a higher-energy state by the energy of photons.

The Reaction Centre

The excitation P680 →P680*of the reaction center pigment P680 occurs here. These special chlorophyll molecules embedded in PS II absorb the energy of photons, with maximal absorption at 680 nm. Electrons within these molecules are promoted to a higher-energy state. This is one of two core processes in photosynthesis, and it occurs with astonishing efficiency (greater than 90%) because, in addition to direct excitation by light at 680 nm, the energy of light first harvested by *antenna proteins* at other wavelengths in the light-harvesting system is also transferred to these special chlorophyll molecules.

This is followed by the step P680* →pheophytin, and then on to plastoquinone which occurs within the reaction center of PS II. High-energy electrons are transferred to plastoquinone. Plastoquinone is then released into the membrane as a mobile electron carrier.

This is the second core process in photosynthesis. The initial stages occur within *picoseconds*, with an efficiency of 100%. The seemingly impossible efficiency is due to the precise positioning of molecules within the reaction center. This is a solid-state process, not a chemical reaction. It occurs within an essentially crystalline environment created by the macromolecular structure of PS II. The usual rules of chemistry (which involve random collisions and random energy distributions) do not apply in solid-state environments.

Link of Water Splitting Complex and Chlorophyll Excitation

When the chlorophyll passes the electron to pheophytin, it obtains an electron from P_{680}^{*}. In turn, P_{680}^{*} can oxidize the Z (or YZ) molecule. Once oxidized, the Z molecule can derive electrons from the water splitting complex.

Summary

PS II is a transmembrane structure found in all chloroplasts. It splits water into electrons, protons and molecular oxygen. The electrons are transferred to plastoquinone, which carries them to a proton pump. Molecular oxygen is released into the atmosphere.

The emergence of such an incredibly complex structure, a macromolecule that converts the energy of sunlight into potentially useful work with efficiencies that are impossible in ordinary experience, seems almost magical at first glance. Thus it is of considerable interest that essentially the same structure is found in *purple bacteria*.

Cytochrome b_6f

PS II and PS I are connected by a transmembrane proton pump, cytochrome b_6f complex (plastoquinol—plastocyanin reductase; EC 1.10.99.1. Electrons from PS II are carried by plastoquinone to *b6f*, where they are removed in a stepwise fashion and transferred to a water-soluble electron carrier called *plastocyanin*. This redox process is coupled to the pumping of four protons across the membrane. The resulting proton gradient (together with the proton gradient produced by the water-splitting complex in PS II) is used to make ATP via ATP synthase.

The similarity in structure and function between cytochrome b_6f (in chloroplasts) and cytochrome *bc*1 (*Complex III* in mitochondria) is striking. Both are transmembrane structures that remove electrons from a mobile, lipid-soluble electron carrier (plastoquinone in chloroplasts; ubiquinone in mitochondria) and transfer them to a mobile, water-soluble electron carrier (plastocyanin in chloroplasts; cytochrome *c* in mitochondria). Both are proton pumps that produce a transmembrane proton gradient.

Photosystem I

PS I accepts electrons from plastocyanin and transfers them either to NADPH (*noncyclic electron transport*) or back to cytochrome b_6f (*cyclic electron transport*)

plastocyanin →P700 →P700* →FNR →NADPH →

- b_6f - plastoquinone

PS I, like PS II, is a complex, highly organized transmembrane structure that contains antenna chlorophylls, a reaction center (P700), phylloquinine, and a number of *iron-sulfur proteins* that serve as intermediate redox carriers.

The light-harvesting system of PS I uses multiple copies of the same transmembrane proteins used by PS II. The energy of absorbed light (in the form of delocalized, high-energy electrons) is funneled into the reaction center, where it excites special chlorophyll molecules (P700, maximum light absorption at 700 nm) to a higher energy level. The process occurs with astonishingly high efficiency.

Electrons are removed from excited chlorophyll molecules and transferred through a series of intermediate carriers to ferredoxin, a water-soluble electron carrier. As in PS II, this is a solid-state process that operates with essentially 100% efficiency.

There are two different pathways of electron transport in PS I. In noncyclic electron transport, ferredoxin carries the electron to the enzyme ferredoxin $NADP^+$ oxidoreductase that reduces $NADP^+$ to NADPH. Alternately, in cyclic electron transport, electrons from ferredoxin are transferred (via plastoquinone) to a proton pump, cytochrome b_6f. They are then returned (via plastocyanin) to P700.

NADPH and ATP are used to synthesize organic molecules from CO_2. The ratio of NADPH to ATP production can be adjusted by adjusting the balance between cyclic and noncyclic electron transport.

It is noteworthy that PS I closely resembles photosynthetic structures found in green sulfur bacteria, just as PS II resembles structures found in purple bacteria.

Photosynthetic Electron Transport Chains in Bacteria

PS II, PS I and cytochromeb6f are found in chloroplasts. All plants and all photosynthetic algae contain chloroplasts, which produce NADPH and ATP by the mechanisms described above. Essentially the same transmembrane structures are also found in cyanobacteria.

Unlike plants and algae, cyanobacteria are prokaryotes. They do not contain chloroplasts. Rather, they bear a striking resemblance to chloroplasts themselves. This suggests that organisms resembling cyanobacteria were the evolutionary precursors of chloroplasts. One imagines primitive eukaryotic cells taking up cyanobacteria as intracellular symbionts.

Cyanobacteria

Cyanobacteria contain structures similar to PS II and PS I in chloroplasts. Their light-harvesting system is different from that found in plants (they use phycobilins, rather than chlorophylls, as antenna pigments), but their electron transport chain

H_2O →PS II →plastoquinone →b_6f →cytochrome c_6 → PSI →ferredoxin →NADPH →b_6f →plastoquinone

is essentially the same as the electron transport chain in chloroplasts. The mobile water-soluble electron carrier is cytochrome c_6 in cyanobacteria, plastocyanin in plants.

Cyanobacteria can also synthesize ATP by oxidative phosphorylation, in the manner of other bacteria. The electron transport chain is

NADH dehydrogenase → plastoquinone → b_6f → cytochrome c_6 → cytochrome aa_3 → O_2

where the mobile electron carriers are plastoquinone and cytochrome *c6*, while the proton pumps are NADH dehydrogenase, b_6f and cytochrome aa_3.

Cyanobacteria are the only bacteria that produce oxygen during photosynthesis. The Earth's primordial atmosphere was anoxic. Organisms like cyanobacteria produced our present-day oxygen containing atmosphere.

The other two major groups of photosynthetic bacteria, purple bacteria and green sulfur bacteria, contain only a single photosystem and do not produce oxygen.

Index

Q

R

S